GENETIC ENGINEERING OF PLANTS AND MICROORGANISMS
IMPORTANT FOR AGRICULTURE

BEP

ADVANCES IN AGRICULTURAL BIOTECHNOLOGY

Akazawa T., et al., eds: The New Frontiers in Plant Biochemistry. 1983. ISBN 90-247-2829-0

Gottschalk W. and Müller H.P., eds: Seed Proteins: Biochemistry, Genetics, Nutritive Value. 1983. ISBN 90-247-2789-8

Marcelle R., Clijsters H. and Van Poucke M., eds: Effects of Stress on Photosynthesis. 1983. ISBN 90-247-2799-5

Veeger C. and Newton W.E., eds: Advances in Nitrogen Fixation Research. 1984. ISBN 90-247-2906-8

Chinoy N.J., ed: The Role of Ascorbic Acid in Growth, Differentiation and Metabolism of Plants. 1984. ISBN 90-247-2908-4

Witcombe J.R. and Erskine W., eds: Genetic Resources and Their Exploitation – Chickpeas, Faba beans and Lentils. 1984. ISBN 90-247-2939-4

Sybesma C., ed: Advances in Photosynthesis Research. Vols. I-IV. 1984. ISBN 90-247-2946-7

Sironval C., and Brouers M., eds: Protochlorophyllide Reduction and Greening. 1984. ISBN 90-247-2954-8

Fuchs Y., and Chalutz E., eds: Ethylene: Biochemical, Physiological and Applied Aspects. 1984. ISBN 90-247-2984-X

Collins G.B., and Petolino J.G., eds: Applications of Genetic Engineering to Crop Improvement. 1984. ISBN 90-247-3084-8

Chapman G.P., and Tarawali S.A., eds: Systems for Cytogenetic Analysis in *Vicia Faba* L. 1984. ISBN 90-247-3089-9

Hardarson G., and Lie T.A., eds: Breeding Legumes for Enhanced Symbiotic Nitrogen Fixation. 1985. ISBN 90-247-3123-2

Magnien E., and De Nettancourt D., eds: Genetic Engineering of Plants and Microorganisms Important for Agriculture. 1985. ISBN 90-247-3131-3

Genetic Engineering of Plants and Microorganisms Important for Agriculture

A seminar held in the framework of the Biomolecular Engineering Programme of the Commission of the European Communities, at the Carlsberg Laboratory in Copenhagen (October 9–10, 1984) with the scientific organization by Professor D. von Wettstein.

Sponsored by the Commission of the European Communities, Directorate-General for Science, Research and Development, Directorate Biology, Radiation Protection and Medical Research, Division Genetics and Biotechnology

edited by

E. MAGNIEN
D. de NETTANCOURT

Division of Genetics and Biotechnology
Commission of the European Communities
Brussels, Belgium

1985 **MARTINUS NIJHOFF/DR W. JUNK PUBLISHERS**
a member of the KLUWER ACADEMIC PUBLISHERS GROUP
DORDRECHT / BOSTON / LANCASTER
for
THE COMMISSION OF THE EUROPEAN COMMUNITIES

Distributors

for the United States and Canada: Kluwer Academic Publishers, 190 Old Derby Street, Hingham, MA 02043, USA
for the UK and Ireland: Kluwer Academic Publishers, MTP Press Limited, Falcon House, Queen Square, Lancaster LA1 1RN, UK
for all other countries: Kluwer Academic Publishers Group, Distribution Center, P.O. Box 322, 3300 AH Dordrecht, The Netherlands

ISBN 90-247-3131-3
EUR 9400 EN

Book information

Publication arranged by: Commission of the European Communities, Directorate-General Information Market and Innovation, Luxembourg

PRINTED IN THE NETHERLANDS

CONTENTS

Other plant-microbe interactions

Plant nuclear genes

Ti/Ri plasmid-derived vectors

Virus-derived vectors

Plant organellar DNA

Plant biochemical mutants and selection procedures

WORKSHOP SESSION ON CYTOPLASMIC MALE STERILITY

WORKSHOP SESSION ON SYMBIOTIC GENES

INTRODUCTION

The previous contractants meeting organized in the framework of the Biomolecular Engineering Programme (BEP) was held in Louvain-la-Neuve in November 1983, approximately one year after the programme was launched. It was in fact a general assembly of all new contractants, bringing together scientists from diverse but complementary backgrounds to try and create what is considered an ideal situation in biotechnology : a melting pot of intermingling disciplines.

One outcome, among many others, of this first scientific event has been the constitution of four coherent working groups, where researchers working on different systems can discuss programmes and results, exchange information, material and staff, and pool their ideas and experiences in the field of biomolecular engineering to achieve specific objectives, such as :

- The development of second generation bioreactors for detoxification or industrial applications;
- The production of substances (vaccines, hormones, ...) for animal husbandry;
- The upgrading of agricultural products (ligno-cellulose, milk, ...) by microbial genetic engineering;
- The improvement of agricultural species by genetic engineering.

The meeting in Copenhagen, 9-10 October 1984, represented the first occasion on which the last of these working groups convened independently. It also provided an opportunity for those who participated in the previous meeting at Louvain-la-Neuve to welcome the new contractants who entered this community programme in April 1984, when the second budgetary phase was activated. Altogether the attendance was composed of 45 project leaders, their co-workers as well as 11 post-doctoral grant recipients, who have built up an expanding community nucleus of advanced research in the field of genetic engineering of plants and their related microorganisms.

The programme of the meeting organized at the Carlsberg Laboratory by D. von Wettstein, was divided into poster sessions, separate workshop sessions chaired by D.R. Davies and J. Tempé, and plenary sessions, where guest scientists (N. Fedoroff, Carnegie Institution of Washington, Baltimore; M.J. Merrick, AFRC unit of nitrogen fixation, Brighton; W.J. Peacock, CSIRO, Canberra; J.D. Rochaix, Université de Genève) and contractants (D. Boulter, C. Elmerich, K. Esser, F. Quétier, J. Schell, O. Schieder) delivered feature lectures as reported below.

The scientific programme was completed by a demonstration session dedicated to Carlsberg facilities and research programmes in the field of basic and applied biology. The demonstrations were commented by the respective project leaders, namely : K. Erdal (brewing from proanthocyanidin-free barley), P. Holm (the synaptonemal complex), S. Holmberg (ILV genes in yeast), K. Breddam (peptide synthesis), U. Rasmussen (screening for barley proteins), P. Sigsgaard (multiferm : screening of the brewing performance of new yeast strains), M. Bohl Pedersen (finger-printing of yeast DNA), G. Høyer Hansen (monoclonal antibodies and thylakoid polypeptides), L. Hallgren (pilot milling and seed analysis), G. Kanangara (chlorophyll biosynthesis).

These proceedings bring together the submissions of the invited speakers, as well as 72 poster contributions, and a summary of the discussions held parallely in the separate workshops, which provide a faithful image of the research carried out within this particular working group. A more detailed presentation of the research results obtained by the contractants, all working groups included, will appear in the 1984 Annual Report of the Biomolecular Engineering Programme, which will be available in April 1985.

While it is not possible to review in the present introduction the numerous achievements reported during the meeting, an attempt is made below to provide a few examples, among several others, of some of the progresses outlined by the contractants on the applications of genetic engineering to plants and microorganisms of agricultural importance.

Nitrogen fixing symbiosis

In the field of symbiotic relationships between plants and microorganisms, most of the results presented at the meeting concerned nitrogen fixing symbiosis between legumes and *Rhizobium* species. They reflected work carried out in five countries of the Community (D, DK, F, IRL, UK) on two essential associations, *R. meliloti* and lucerne on the one hand, *R. leguminosarum* and pea or field bean on the other hand. This work is being concentrated at 4 different levels, corresponding to 4 sets of genes considered necessary to the establishment of an efficient symbiosis, namely nitrogen fixation genes (Fix), nodulation genes (Nod), competition genes, all those belonging to the microbial partner, and finally a few genes from the plant partner which are specifically activated in relation with the establishment of the symbiosis (leghemoglobin or glutamine synthetase genes). As concerns the genes from *Rhizobium*, many contributions dealt with their localization on a special symbiotic plasmid, their molecular structure and specific functions, in an effort which brings us to a stage of advanced understanding of the genetics of nitrogen fixing symbiosis.

In particular, the statements were made that many Fix genes, although they most probably correspond to similar functions in different symbionts, bear few sequence homologies from one species to another, with some exceptions where R. meliloti probes could help in identifying R. leguminosarum homologous genes (Manian and Pühler)*. Most interesting is the discovery by the group of Boistard et Dénarié (Huguet et al.) that Fix genes can be located far (200 Kb) from the main Fix region of the symbiotic plasmid in R. meliloti. Nodulation genes (Nod) are clustered, they are few in number (less than 10Kb required for nodulation ability) and look similar in different species (Downie et al.). The identification of genes determining host-range is an essential issue in that work, one of its objectives being to broaden the spectrum of associations that a particular nitrogen fixing Rhizobium can establish with crop species. The degree of specificity of these associations is very high within the legume family, but researchers of the same group (Götz et al.) brought evidences that the manipulation of a small region of the Nod cluster in the symbiotic plasmid of R. leguminosarum is sufficient to extend the nodulation ability of this symbiont from western varieties of peas to a recalcitrant cultivar originating from Afghanistan.

In this field, research is obviously far from bringing about the clues which will lead to super-productive engineered symbioses in agricultural species, but it is important to stress that each step made in the molecular characterization of the system represents a progress towards unravelling the intricate puzzle of plasmids and genes which operate in such a variety of different interacting species. This is typically an area of labour intensive research requiring the joint effort of many laboratories to encompass a significant number of systems and species between which probes are exchanged and methodologies are comparatively assessed, an effort which appears particularly profitable in the european framework as shown here.

Other symbioses in the soil

In the field of symbiosis in general, other beneficial interactions which might profit to plant growth, research is by far at a much more preliminary stage.

However, the potentialities of these symbioses might be extremely high. Two communications reporting work at an early stage of advancement should be mentioned : the discovery of plasmids in ectomycorrhiza such as Morchella (Meinhardt and Esser) and the characterization of Pseudomonas putida mutant strains with different iron-binding capacities (Marugg et al.), are opening the way to the understanding of other associations in the soil, which prove to have significant plant growth stimulating properties.

* All references made in this introduction apply to the abstracts presented in this volume.

As far as symbioses are concerned, a theoretical distinction can be made between mutualistic and antagonistic symbioses. And in certain cases indeed, the transition between real symbiosis and parasitism might appear weak, considering certain common features of the early steps of recognition, root colonization and communication between plants and microbes. An horizontal approach to symbiosis in the broad sense is made even easier from the point of view of molecular genetics. Fourty contractants having experience in the field of plant-microbe interaction met at a workshop session and discussed some implications of this conceptual approach. We were first reminded that Agrobacterium itself, well known for its ability to transform wounded plant tissues in nature, is an endorhyzobacteria which can be detected in the radicular system of corn plants. It appeared rapidly that although the mechanism for transfer and integration of the T-DNA into the plant host chromosomal DNA were being described in a more refined way than ever, our knowledge was far from being sufficient on the virulence functions, also coded on the Ti plasmid of Agrobacterium. A contracting laboratory at Leiden has isolated a compound which can induce Vir (virulence) genes, even in the absence of wounded plants or cultured cells. In the same way that rhizobiologists focus their attention on signals exchanged between host-plant and Rhizobium to activate sets of genes in relation with the establishment of symbiosis, agrobacteriologists also acknowledge the need for an increased attention on those functions, partly coded from the Vir region of the Ti plasmid, which control the infection of the plant by Agrobacterium.

Particularly interesting is the finding by the group of J. Tempé of a striking analogy between Agrobacterium and Rhizobium, showing that nodules produced by a particular R. meliloti strain contain a substance which fits all criteria for an opine-like substance (normally restricted to crown-gall systems), that is a nodule-specific compound, strain-specific, degradated by the interacting microorganisms. In the same way that the Ti plasmid of Agrobacterium contains all catabolic functions for octopine or nopaline degradation, Tempé demonstrated that genes for the degradation of this new opine-like substance were present on the symbiotic plasmid of Rhizobium.

Agrobacterium as a system for gene transfer to plants.

One of the most impressive presentations, by M. Van Montagu, concerns the report of a monocot being transformed by the Ti-plasmid system : this was achieved in Asparagus, where the transformed tissue has the ability of hormone independent growth like normal crown-gall, and of producing nopaline. The laboratory of Prof. Schilperoort, in Leiden, reached similar results with other monocots such as onion or daffodils, showing by these repeated successes that there is now scope for extended applications of Ti-plasmid-derived vectors in more numerous crop species than initially expected. A transcription system has also been made functional in another monocot : J. Schell reported on a black mexican maize protoplast line, with high plating efficiencies, which took up, through liposome-delivery, a chimeric construct made of the nos (nopaline synthase) promotor of the T-DNA and the bacterial gene for chloramphenicol transferase.

There has been a clear transient expression of the nos promotor, which definitely establishes that the promotor is also functional in monocots. On the other hand, the reverse experiment designed for obtaining the expression of a promotor of a monocot (Adh1 gene from maize) with a variety of constructs in tobacco as a receptor invariably failed, although the heterologous expression of such a promotor could be obtained in animal systems (Peacock). J. Schell further described the recent success of his multipolar group (Gent-Brussels-Köln), in achieving the regulated expression of chloramphenicol transferase gene under the control of the light-sensitive promotor of the small subunit of ribulose biphosphate carboxylase (ss RBCS). The expression is strictly controlled by light and is possible only in plant tissue with an active phytochrome system. Searching for sequences which control tissue specific expression, this group also isolated from cDNA banks of leaf, root and tubers of potato 3 different leaf-specific genes, which were shown to be actually light inducible.

The main problems associated with monofactorial gene transfer experiments in plants obviously remain at 3 critical levels : finding delivery systems amenable to the monocots, regenerating plants from engineered single cells and controlling the expression of engineered genes in their heterologous environment. The results presented in Copenhagen show that a solution is being approached in the most rational way through a tedious study of regulatory DNA sequences within chimeric genes, in well chosen transformation and/or transcription systems. However brilliant the above results might be, we are still at a stage where more questions are being raised than conclusive answers produced.

The amount of effort which is continuously invested in a coordinated manner within a transnational group such as that of J. Schell for instance, and finally between the different groups constituted in the framework of this programme, illustrates the complexity of the problem.

Other transformation systems

It is possible, in the absence of a systematic approach to the infection of monocots by *Agrobacterium*, to envisage other methods of gene transfer. D. Boulter illustrated feasibility experiments on microinjection of DNA into germinating pollen tubes or into ovaries (Hepher et al.), which allow gene transfer into monocots to be realistically contemplated without having to rely upon the as yet inapplicable *Agrobacterium* system. With regard to the transformation of certain dicots, which are not as yet amenable to protoplast culture and regeneration, J. Schell and M. Van Montagu demonstrated the possibility to transform such species by direct double infection on the wounded stem of a plant : one Ti-plasmid with cytokinin genes only, to force direct shoot regeneration from the crown-gall tissue, and another Ti-plasmid carrying the foreign gene.

Other progresses

Many other reports at more preliminary stages included research on the isolation and characterization of plant genes. Significant results were obtained in molecular studies on the structure and function of storage protein genes from cereals or legumes (Croy et al.; Casey et al.; Kreis et al.; Thompson et al.; Rasmussen et al.; Viotti et al.) and of mitochondrial genomes (Lejeune et al.; Goblet et al.; Van der Mark and Kool; Mezzarobba et al.; Davies and Powling; Briquet et al.; Charbonnier et al.). The molecular organization of wheat mitochondrial DNA and of its multiple recombinational forms has been largely elucidated by F. Quêtier (Quêtier and Lejeune; Lejeune et al.), which is an important step towards future possibilities of manipulating mitochondrial properties (resistances to herbicides or expression of male sterility). Refinements of the protoplast fusion methodologies are being worked out to select and stabilize new hybrid combinations between naturally incompatible species. These include the adaptation to plant protoplasts of electro-induced fusion (Tempelaar and Jones; Sotak et al.), the development of molecular probes to characterize fusion products (Sala et al.; Müller-Gensert et al.), the obtention of asymetric hybrids by radiation inactivation of one progenitor (Schieder; Foulger et al.; Famelaer et al.).

Transnational cooperation

The meeting strongly intensified transnational cooperations which had been established at earlier meetings. Among many examples, one could mention multiple cooperation programmes between Orsay, Harpenden, Norwich and Rome which foresee the exchanges of strains and plasmids; laboratories at Amsterdam and Orsay are to join the integrated group Norwich, Louvain-la-Neuve and Dijon, dedicated to the study of the molecular basis of male-sterility (Kool, Quêtier, Davies, Briquet, Cornu). A new type of plant biochemical mutant (Negrutiu and Jacobs) will be sent from Brussels to Pavia (Sala) to be used in somatic hybridization experiments between carrot and tobacco, following a scheme which was discussed at the meeting between these groups.

In line with the original intentions of the Commission, the visible outcome of the meeting in Copenhagen appears to go far beyond what one might just consider another scientific event in a very busy sector. It really shows that a European strategic strength is being built up in the field of agricultural biotechnology.

ACKNOWLEDGEMENTS

The services of the Commission express their gratitude to Carlsberg Laboratory at Copenhagen for having kindly accepted to host this meeting of contractants working in the framework of the Biomolecular Engineering Programme (BEP) of the C.E.C.
They wish to thank the Carlsberg Breweries for their contribution to the social programme.
Very special thanks are due to Professor D. von Wettstein, as the excellent organizer of the scientific programme. His invaluable contribution to many organizational steps and his constant cooperation during all stages of the preparation of the meeting are deeply acknowledged.

SCIENTIFIC PROGRAMME

Tuesday, Oct. 9

08:00-09:30	Setting up of posters
09:30-10:30	Dr. W.J. Peacock, CSIRO, Canberra: Regulation of gene expression in plants
10:30-11:30	Dr. J. Schell, Max-Planck-Inst. f. Züchtungsf., Köln: Expression of introduced genes in higher plants
11:30-13:00	Poster viewing
13:00-14:00	Lunch
14:00-14:45	Dr. O. Schieder, Max-Planck-Inst. f. Züchtungsf., Köln: Genetic manipulations by protoplast fusion
14:45-15:30	Dr. J.D. Rochaix, University of Geneva: *Chlamydomonas reinhardii*, a potential model system for chloroplast gene manipulation
15:30-16:15	Dr. N. Fedoroff, Carnegie Institution, Baltimore: Transposing elements in maize
16:15-17:30	Poster viewing and refreshments
17:30-19:00	Demonstrations of the basic and applied research at the Carlsberg Research Centre in groups of 10
19:30	Dinner and entertainment by invitation of the Carlsberg Breweries

Wednesday, Oct. 10

09:00-13:00	Poster viewing

Parallel sessions

09:00-12:00	Workshop on cytoplasmic male sterility (organized by Prof. D.R. Davies)
09:00-12:00	Workshop on symbiontic genes (organized by Dr. J. Tempé)
13:00-14:00	Lunch
14:00-14:10	Dr. D. Boulter, University of Durham: Microinjection of DNA into pollen ovules and somatic cells
14:20-14:40	Discussion
14:40-15:00	Dr. C. Elmerich, Institut Pasteur, Paris: Nitrogen fixation
15:00-15:20	Discussion
15:20-15:40	Dr. M.J. Merrick, ARC Unit of Nitrogen Fixation, Brighton: The *nif* genes and their transfer into plant cells
15:40-16:00	Discussion
16:00-16:20	Dr. K. Esser, University of Bochum: Molecular cloning with mitochondrial DNA
16:20-16:40	Dr. F. Quétier, Université Paris, Orsay: Recombination of mitochondrial DNA
16:40-17:00	Discussion
17:00	Goodbye

PLENARY LECTURES

ALCOHOL DEHYDROGENASE GENES AND CONTROLLING ELEMENTS IN MAIZE

W.J. PEACOCK, E.S. DENNIS, W.L. GERLACH, J. ELLIS, D. LLEWELLYN & J. WALDRON

Division of Plant Industry, CSIRO, Canberra, Australia

Anaerobic conditions induce expression of the two alcohol dehydrogenase genes of maize. We have isolated and sequenced the *Adhl* gene from chromosome 1 and the *Adh2* gene from chromosome 4 (1,2,3). Time course analyses of steady state mRNA levels and of run-off transcription in isolated nuclei have shown that the induction of gene expression is at a transcriptional response level.

Both genes, which have evolved from a common ancestral gene, have nine intervening sequences. Their coding regions have 82% sequence homology but sequences have diverged in the introns and in the 3' and 5' flanking sequences. There are several short segments of sequence conservation in the 5' transcribed and non-transcribed regions which may be important control sequences for the co-ordinate expression of the genes.

In an attempt to carry out a functional analysis of the putative control sequences the *Adhl* promoter was spliced to the octopine synthase cassette of *Agrobacterium* and introduced into *Nicotiana* sp. via the T-DNA system. The promoter was not functional in tumour cells, nor was it functional when the whole *Adhl* genomic segment was transformed into *N. tabaccum* or *N. plumbaginifolia* and activity assayed in both tumour cells and regenerated plants.

Expression of *Adhl* is severely modified in the *Ds* controlling element mutant, *Adhl-Fm335* (4). We have shown that the *Ds* element, 405 bp in length, is inserted between nucleotides 45-46 in the 5' transcribed/-non-translated region of the gene (5). Run-off transcription experiments show the mutant gene is anaerobically induced and has a time course of transcription rates comparable to the progenitor gene. The steady state level of mRNA is two orders of magnitude less than that for the progenitor allele. S_1 mapping has established that the mutant uses the same transcription start point and that a large component of the Ds insert (+25-35 → 405) is processed as a tenth intervening sequence in the transcript. The intervening sequence acceptor sequence is a good fit to the consensus but the donor sequence is imperfect and may provide an explanation for the inefficiency in processing of the transcript.

A number of revertants of the mutant, induced by the action of the autonomous element Ac, show the Ds insertion to have been excised. Each revertant retains the 8 bp repeat generated by the original Ds insertion but with nucleotide deletions and/or transversions at the new junction.

We have made a variety of plasmid constructs with the Adh1 gene and the Ds and Ac elements, and attempted to introduce these sequences into the embryo sac of maize via the pollen tube. Techniques for monitoring DNA entry into the pollen tube and for achieving effective pollination with pre-germinated pollen have been developed, but we have no confirmed cases of DNA transformation among the ca. 2000 embryos examined.

References

1. Gerlach, W.L., Pryor, A.J., Dennis, E.S., Ferl, R.J., Sachs, M.M. and Peacock, W.J. (1982) Proc.Natl.Acad.Sci.USA, 79, 2981-2985.
2. Dennis, E.S., Gerlach, W.L., Pryor, A.J., Bennetzen, J.L., Inglis, A., Llewellyn, D., Sachs, M.M., Ferl, R.J., Peacock, W.J. (1984) Nucl.Acids Res., 12, 3983-4000.
3. Dennis, E.S., Sachs, M.M., Gerlach, W.L., Finnegan, E.J. and Peacock, W.J. (1984) Nucl.Acids Res., submitted.
4. Osterman, J.C. and Schwartz, D. (1981) Genetics, 99, 267.
5. Sutton, W.D., Gerlach, W.L., Schwartz, D. and Peacock, W.J. (1984) Science, 223, 1265-1268.

REGULATION OF EXPRESSION OF GENES INTRODUCED IN PLANTS WITH TI PLASMID VECTORS

Jeff Schell[*+], Marc van Montagu[*], Patty Zambryski[*], Luis Herrera-Estrella[*], Lothar Willmitzer[+], Fritz Kreuzaler[+], Ann Depicker[*], Peter Schreier[+], Hans Bohnert[°].

[+]Max-Planck-Institut für Züchtungsforschung, 5000 Köln 30, FRG;
[*]Laboratorium voor Genetika, Rijksuniversiteit Gent, 9000 Gent, Belgium;
°University of Arizona, Dept. of Biochemistry, Tucson, Arizona 85721, USA.

Large plasmids in _Agrobacterium tumefaciens_ (Ti) and _A. rhizogenes_ are known to endow these bacteria with the capacity to transfer a defined DNA fragment (T-DNA) into plant cell nucleus and to covalently integrate this T-DNA segment in chromosomal DNA, thus creating a new locus at a number of possible sites. Under normal circumstances only the T-region of the Ti-plasmid is inserted in the chromosomal DNA and thus stably maintained. No functions located within the T-region are required for either transfer or integration. It was possible to design modified Ti-plasmids that can insert foreign genes in plant cells from which normal plants can be regenerated that express the foreign genes and transmit them sexually with normal mendelian segregation ratios. Recent work has indicated that small (+/- 25 bp) so called "Integration sequences" bordering the T-DNA region of Ti-plasmids are responsible and essential for the oriented transfer and for integration of foreign genes covalently linked to this integration sequences. Probably because of the precision of this transfer/integration system DNA introduced in plants with Ti-plasmid vectors is usually integrated without major rearrangements which is in contrast with transformation based on naked DNA uptake by plant protoplasts. Several foreign genes thus introduced in plants failed to be expressed. To test whether or not 5'upstream sequences of genes that are expressed in plants can be used to promote the expression of coding sequences of foreign genes, chimeric genes were constructed and introduced in plants. General promotor regions were used

1° the 5'promotor region of the nopaline system gene
2° the 5'promotor region of the small subunit of RuBisCo
3° the 5'promotor region of the chalcone synthase.

Different coding sequences were fused to these promotor sequences. Thus it was shown that the nopaline synthase promotor allowed the constitutive expression of neomycin-phosphotransferase, of chloramphenicol trans-acetylase and of a Mtx resistant dihydrofolate-reductase of E.coli. The RuBisCo promotor was shown to allow the light inducible expression of a chloramphenicol transacetylase and of neomycine-phosphotransferase. These experiments open the way for a very detailed analysis of the regulation of gene expression in plants. Similar experiments involving the use of chimeric genes constructed in vitro and introduced back in plants are being conducted to study the light regulation of the Chalcone synthase gene to study the mechanisms underlying organ specific expression in potatoes. Furthermore it was demonstrated that it is possible to transport foreign proteins in plastids provided they are linked to a plastid specific transit peptide sequence. Chimeric genes coding for fusion proteins consisting e.g. of the transit peptide of the small subunit of RuBisCo fused to the E.coli neomycine-phosphotransferase were shown after introduction in tobacco, to yield kanamycin resistant plants. Chloroplasts from such plants were shown to contain the active neomycine-phosphotransferase enzyme.

GENETIC MANIPULATION BY PROTOPLAST FUSION

O. SCHIEDER, Max-Planck-Institut für Züchtungsforschung, D-5000 Köln 30, Fed. Rep. of Germany

1. INTRODUCTION

The asexual combination of two different lines or species by protoplast fusion opens the possibility to produce hybrids which are not obtainable by sexual methods. Apart from the production of somatic hybrids where the complete genomes of two fusion partners are combined, protoplast fusion can also be used for the transfer of only a few chromosomes or even smaller nuclear genetic units.

2. SOMATIC HYBRIDIZATION

Complete genomes were combined by using protoplasts of albino mutants of *Datura innoxia* and green protoplasts of four other *Datura* species that were unable to develop. Green and developing cell colonies were selected as somatic hybrids. Both highly fertile and sterile tumorous amphidiploid somatic hybrids were obtained (1).

3. CHROMOSOME ELIMINATION

Transfer of only a few chromosomes was achieved by fusing albino protoplasts of *D. innoxia* with green wild-type protoplasts of *Atropa belladonna*. In the resulting somatic hybrids, certain *A. belladonna* chromosomes were eliminated, as was revealed by the appearance of either albino sectors or cytological studies. Simultaneously with the elimination of the *A. belladonna* chromosomes redifferentiation of shoots was obtained. In several stable green sublines only 2 or 4 *A. belladonna* chromosomes instead of 72 were observed (1).

4. TRANSFORMATION EXPERIMENTS

For the transfer of genes, wild-type protoplasts of *Physalis minima* and *D. innoxia* were inactivated by X-rays and fused separately with protoplasts of a NR^- (CNX-68) mutant of *Nicotiana tabacum*. On medium lacking amino acids but containing NO_3^- as the nitrogen source, several cell lines were selected in which NR activity was restored (1).

In another interspecific reconstruction experiment, the albino character of a nuclear mutant of *D. innoxia* was restored after fusion with inactivated wild-type protoplasts of *P. minima*. The transformation of *Physalis* genes in *Datura* was also confirmed by an increase in the amount of nuclear DNA, the observations of few chromosomes from *P. minima*, and the presence of donor isoenzymes (2).

In a third experiment protoplasts of a cell line of *N. paniculata* - transformed by *Agrobacterium tumefaciens* B6S3 - were isolated, inactivated by X-rays, and fused with protoplasts of a NR^- cell line of *N. tabacum* (CNX-68). After 4 to 6 weeks of cultivation in a medium containing plant hormones and amino acids, the developed cell colonies were transferred onto a medium lacking plant hormones but supplemented with $KClO_3$. By this treatment only colonies could survive which were deficient in NR but contained the T-DNA of B6S3. Until now, 6 cell lines were selected synthesizing octopine, showing hormone independent growth, and lacking any NR activity. Moreover, the selected cell lines exhibited an additional recessive marker. The cell line CNX-68, used as a recipient, is possessing a Ca-oxalate druse which disappears in symmetric somatic hybrids between CNX-68 and *N. paniculata*. This Ca-oxalate druse is present in all the selected transformed cell lines (3).

REFERENCES

1. Schieder O, Gupta PP, Krumbiegel G, Hein T. 1983. Protoplast fusion and transformation. In: Better crops for food. Ciba Foundation Symposium No. 97, Pitman, London, pp. 213-227.
2. Gupta PP, Schieder O, Gupta M. 1984. Mol. Gen. Genet. in press.
3. Unpublished results.

CHLOROPLAST MOLECULAR GENETICS AND TRANSFORMATION IN CHLAMYDOMONAS REINHARDII.

J.D. ROCHAIX, J. ERICKSON, M. GOLDSCHMIDT-CLERMONT, M. HERZ, R.J. SPREITZER AND J.M. VALLET.

Departments of Molecular Biology and Plant Biology, University of Geneva, CH-1211 Geneva 4, Switzerland.

The green, unicellular alga *Chlamydomonas reinhardii* represents an attractive model system for a combined genetic and molecular analysis of the structure, function and expression of genes involved in photosynthesis. We have examined several mutations at the nucleotide level in the two chloroplast genes *rbc*L and *psb*A that code for the large subunit of ribulose 1,5 bisphosphate carboxylase/oxygenase (Rubisco) and the 32 Kdalton thylakoid polypeptide, respectively.

While the first Rubisco mutant examined carried a missense mutation (1), two other mutants (2) contain nonsens codons in their *rbc*L genes. Pulse-labelling and pulse-chase experiments reveal that although these two mutants do not produce Rubisco holoenzyme the small subunit is made and apparently imported into the chloroplast where it is rapidly degraded.

The *psb*A gene product has been identified as the herbicide-binding protein (3). In *C. reinhardii* the gene spans a region of nearly 7kb because of the presence of four introns. Since *psb*A is located within the chloroplast inverted repeat, it is present in two copies per chloroplast genome (4). In collaboration with P. Bennoun (5) and L. Mets (6), we have examined *psb*A from 3 uniparental mutants resistant to diuron and/or atrazine. These mutants have single base pair changes which result in amino acid substitutions at residues 264 (5), 219 and 255 (unpublished results) indicating that there is a considerable flexibility in the binding sites for these herbicides. The *psb*A gene appears to be particularly prone to suffer rearrangements since among 6 photosystem II deficient, uniparental mutants examined, all have both *psbA* genes deleted.

The molecular characterization of these mutants opens the door for serious attempts at chloroplast transformation. We have been able to successfully transform *C. reinhardii* at the nuclear level (7) and it has been possible to isolate DNA fragments that promote autonomous replication in *C. reinhardii*; presumably in the nucleocytoplasmic compartment (8). Surprisingly, these ARC elements hybridize exclusively with chloroplast DNA and one of them maps very near an authentic chloroplast origin of DNA replication (9).

REFERENCES

1. Dron, M, Rahire, M, Rochaix, JD, and Mets, L. 1983. Plasmid, 9, 321-324.
2. Spreitzer, RJ, and Ogren, W. 1983. Proc. Natl. Acad. Sci. USA 80, 6293-6297.
3. Pfister, K, Steinback, KE, Gardner, G, and Arntzen, CJ. 1981. Proc. Natl. Acad. Sci. USA 78, 881-885.
4. Erickson, J, Rahire, M, and Rochaix, JD. 1984. submitted
5. Erickson, J, Rahire, M, Bennoun, P, Delepelaire, P, Diner, B, and Rochaix, JD. 1984. Proc. Natl. Acad. Sci. USA 81, 3617-3621.
6. Galloway, R, and Mets, L. 1984. Plant Physiol. 74, 469-474.
7. Rochaix, JD, and von Dillewijn, J. 1982. Nature, 296, 70-72.
8. Rochaix, JD, Rahire, M, and von Dillewijn, J. 1984. Cell 36, 925-931.
9. Waddell, JM, Wang, XM, and Wu, M. 1984. Nucl. Acids. Res. 12, 3843-3856.

Supported by grant 3.258.0.82 from the Swiss National Science Foundation

MOLECULAR STUDIES ON THE TRANSPOSABLE CONTROLLING ELEMENTS OF MAIZE

N.V. FEDOROFF

1. INTRODUCTION

The transposable controlling elements of maize were identified and analyzed genetically by B. McClintock beginning in the early 1940's. The first elements whose transposition was detected were designated the Dissociation (Ds) and Activator (Ac) loci for their ability to provide a site for and activate the process of chromosome breakage, respectively. Another family of transposable elements studied extensively by McClintock is the Suppressor-mutator (Spm) family. We have recently isolated elements belonging to both the Ac-Ds and Spm controlling element families (1-4).

2. RESULTS

2.1 Ac and Ds elements. An Ac and two different Ds elements were cloned from strains with insertion mutations at the *waxy* (*wx*) locus. The Ac element is 4653 nucleotides long and contains two major open reading frames (ORFs). The ORFs diverge from an intergenic region which contains eukaryotic transcription initiation sequences and are followed by polyadenylation sites. The larger of the two ORFs is believed to encode the element's transposase, because a Ds element directly derived from the original Ac element differs from it by a 194 bp deletion in the large ORF and can only transpose in the presence of an Ac element. A small Ds element of 2 kb was also isolated and comprises the ends of Ac, lacking almost all of the large ORF. The Ac and large Ds elements both terminate in the same imperfect 11 base pair inverted repetition and generate an 8 bp direct duplication upon insertion. A single *Wx* revertant has been examined and retains 6 bp of the 8 bp duplication generated upon insertion.

2.2 Spm elements. Although we have isolated a defective Spm element from the *wx* locus, most loci with non-defective Spm elements encode proteins involved in anthocyanin biosynthesis, few of which have been identified. Moreover, the known anthocyanin biosynthetic enzymes are made in very small amounts. As a consequence, we adopted an indirect strategy for cloning Spm elements, first isolating the bronze (bz) locus by 'transposon tagging' and then cloning bz alleles with Spm insertion mutations.

2.2.1 Cloning the bz locus. The bz locus encodes an anthocyanin biosynthetic enzyme and mutant alleles have been isolated which have Ac insertions, as well as an insertion of a non-defective Spm element. A subcloned internal fragment of the Ac element was used initially as a probe to identify homologous cloned genomic DNA fragments containing Ac-like sequences. Sequences flanking the Ac elements were then subcloned and tested for homology to the bz locus by using them to probe genomic DNA from strains with well-characterized insertion and deletion mutations at the bz locus. A fragment with the expected properties was identified and used to clone a non-mutant Bz allele of the locus, as well as alleles with Spm insertion mutations (3).

2.2.2 Spm elements from the wx and bz loci. Genomic fragments were cloned from the bz-m11 and bz-m13 alleles (isolated by O. Nelson) which behave genetically as insertions of a non-defective Spm and a defective Spm element, respectively. Both alleles have insertions of 2.2-2.5 kb. The element present in the bz-m13 allele is indistinguishable from the defective Spm element cloned from the wx locus. The insertion in the bz-m11 allele is markedly different. We have cloned a 1.5 kb insertion which appears to correspond to the ends of the element. The insertion cloned from the bz-m11 and bz-m13 alleles do not cross-hybridize.

REFERENCES

1. Fedoroff, N., Wessler, S., and Shure, S. 1983. Isolation of the transposable maize controlling elements Ac and Ds. Cell 35, 243-251.
2. Pohlman, R.F., Fedoroff, N.V., and Messing, J. 1984. The nucleotide sequence of the maize controlling element Activator. Cell 37, 635-643.
3. Fedoroff, N., Furtek, D., and Nelson, O. 1984. Cloning of the bronze locus in maize by a simple and generalizable procedure using the transposable controlling element Ac. Proc. Nat. Acad. Sci. USA 81, 3825-3829.
4. Fedoroff, N., Shure, M., Kelly, S., Johns, M., Furtek, D., Schiefelbein, J., and Nelson, O. 1984. Isolation of Spm controlling elements from maize. Cold Spring Harbor Symp. Quant. Biol., in press.

Microinjection of DNA into pollen, ovaries and somatic cells

Hepher A., Gates P. & Boulter D.

Genetic transformation via pollen vectors has several potential advantages in the generation of new crop varieties over both conventional plant breeding and more novel techniques such as the use of plasmid vectors. Apart from DNA extraction, manipulation and injection it follows standard plant breeding methodology. However only one or a few genes may be transferred, so reducing the number of plant generations required to stabilise the new variety, and sterility barriers may be overcome, permitting inter-generic gene transfer. In contrast to plasmid vectors the need for embryo culture or regeneration of plants from callus is eliminated and monocotyledonous plants are not excluded. Additionally many valuable agronomic traits such as yield and stress resistance are probably controlled by gene complexes which may be 'scattered' over the genome; pollen mediated transformation using sub-sets of total nuclear DNA may be a possible approach to this problem.

The introduction of small volumes (typically 10^{-11} ml) of various substances by direct microinjection is a well established technique in zoological research (DIACUMAKOS, 1973) and stable transformation rates of 20-30% have been reported after intranuclear injection of DNA (CAPECCHI, 1980; GRAESSMAN, GRAESSMAN and MUELLER, 1981). However, problems with the penetration of the cell wall by microcapillaries have meant that the botanical application of this technique has centred principally upon injections into protoplasts (STEINBISS and STABEL, 1983; GRIESBACH, 1983).

The basic requirements for microinjection are an inverted microscope with long working distance objective lenses and a micromanipulator with both rough and fine 3-dimensional control, mounted on a vibration-free bench. Microcapillary tubes with an external tip diameter of less than 1 µm, containing an inner filament to allow back-filling, may be produced by a variety of specially designed pullers. Substances may be injected either by iontophoresis or, more usually for larger molecules, by use of compressed air or nitrogen. Injection rates in the system at Durham are controlled by a series of electrically operated valves (modified from ANSORGE, 1983), giving an injection pressure and a purge pressure, together with a holding pressure to reduce suck-back and subsequent capillary blockage.

The thin cell walls of the rapidly growing pollen tube may easily be pierced by microcapillaries with tips less than 1 µm diameter and pollen germinated *in vitro* may be transferred to the stigma and will continue to grow in a manner comparable with pollen germinated *in situ*. Pollen nuclei are generally not visible using bright field or interference optics but may be visualised by incorporation of a non-toxic DNA-specific fluorochrome such as 4'-6 diamidino-2 phenylindole into the germination medium.

The DNA to be injected should contain the desired foreign gene, a screenable, ideally selectable marker, and possibly transposition sequences to increase the chance of incorporation into the host genome. Equally the choice of recipient should be carefully considered. Pollen should germinate easily and consistently *in vitro*, producing relatively large tubes and, for intranuclear injection, large nuclei. In addition the floral anatomy should be amenable to manipulation.

Recent work carried out with *Zea mays* has suggested that extra-cellularly applied DNA can be taken up by germinating pollen and stably incorporated. This in conjunction with reports that characters specific to DNA injected into the placental tissue of cotton during fertilisation are exhibited in the subsequent progeny (ZHOU et al. 1983) indicates that in the future this may provide a satisfactory method. However, so far as we are aware, it has so far not been 'reduced to practice'.

References

Ansorge W. (1982) "Improved system for capillary microinjection into living cells". Exp. Cell. Res. 140, 31-37.

Capecchi M.R. (1980) "High efficiency transformation by direct micro-injection of DNA into cultured mammalian cells". Cell 22, 479-488.

Diacumakos E.G. (1973) Methods for micromanipulation of human somatic cells in culture. In: Methods in Cell Biology 7 (ed. Prescott D.M.), 287-311.

Griesbach R.J. (1983) Protoplast microinjection. Plant Molecular Biol. Reporter 1 (4), 32-37.

Graesman A., Graessman M. & Mueller C. (1981) Regulation of SV40 gene expression. Adv. in Cancer Res. 35, 111-149.

Steinbiss H.-H. & Stabel P. (1983) Protoplast derived tobacco cells can survive capillary microinjection of the fluorescent dye lucifer yellow. Protoplasma 116, 223-227.

Zhou G.-Y, Weng J., Zeng Y., Huang J., Qian S. & Liu G. (1983) Introduction of exogeneous DNA into cotton embryos. In : Methods in Enzymology 101 (Academic Press), 433-481.

NITROGEN FIXATION

C. ELMERICH

INTRODUCTION

Only some prokaryotes have evolved the genetic information for nitrogen fixation, ie. utilization of gazeous nitrogen as N-source. These prokaryotes, when associated with plants, can contribute to their growth by providing reduced nitrogen (ammonia).

THE NITROGEN FIXATION (*NIF*) GENES

Most of what is known about the genetics and regulation of nitrogen fixation was initially established in *Klebsiella pneumoniae* M5a1. In this bacterium a cluster of 17 *nif* genes organized in 7 or 8 transcription units has been identified (1) (see Figure). The structural genes for the nitrogenase polypeptides are *nifK* and *nifD* for the MoFe protein and *nifH* for the Fe protein. Molybdenum is associated with the MoFe protein through a cofactor, the FeMoco which plays a role in catalysis. Five genes, *nifQ,B,V,N,E*, appear to be essential for formation of a functional MoFe protein and *nifM* and *nifS* are thought to be essential in the processing of the Fe protein. However, little is known about the biochemical functions and properties of the corresponding products. *NifF* and *nifJ* code for the components of the e^- transport chain. No function has as yet been found for *nifX,U,Y* nor for the open reading frame (ORF) sequence preceeding *nifJ*. *NifLA* are involved in the regulation of *nif* expression, *nifA* coding for an activator and *nifL* for a repressor (2).

The *nifHDK* genes are highly conserved in nitrogen-fixing bacteria. *K. pneumoniae* probes were used to screen for homology and to clone the corresponding DNA of many other diazotrophs including several *Rhizobium* species, cyanobacteria, photosynthetic bacteria, *Azotobacter* and *Azospirillum*. In these genera, data on the identification and the organization of the *nif* genes are still limited.

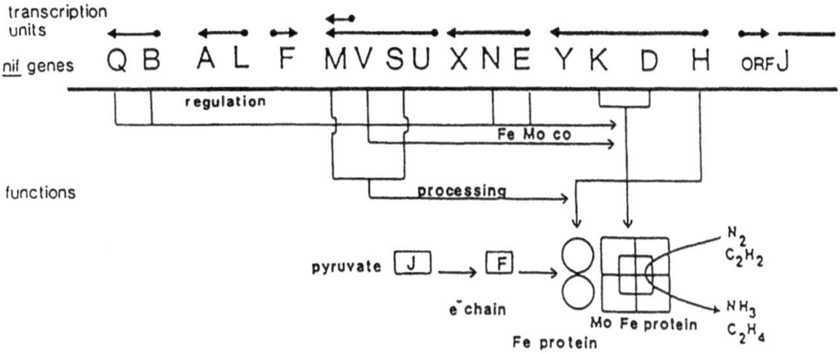

SYMBIOSIS AND ASSOCIATION WITH PLANTS

The Rhizobium-legume symbiosis, which is responsible for about 30% ($40x10^6$ tons) of the total nitrogen fixed, is the most documented symbiotic system at the molecular level (3). Rapid progress is occuring in the identification of the bacterial genes involved in the process. An important feature that greatly facilitated this analysis was the discovery that symbiotic genes were plasmid-borne in several strains of Rhizobia.

Research on nitrogen fixation with non-leguminous plants is still at the ecological stage. However, major interest is focussed on actinorhizal plants. These plants carry root nodules infected with an actinomycete of the Frankia genus and can have important ecological and agronomic applications in reforestation and biomass production (4).

Various diazotrophs e.g. Azospirillum spp., Azotobacter spp., Enterobacteriaceae proliferate in the root zone of grasses (5) . Their role in increasing crop productivity, apart a few exceptions, is low if not negligible. However, these bacteria appear as potential fertilizers and the question as to engineer a real bacteria-grass symbiosis is open.

REFERENCES

1. Robson R, Kennedy C & Postgate JR. 1983. Can. J. Microbiol. 29 : 954-967.
2. Merrick MJ. This volume.
3. Stacey G & Upchurch RG. 1984. Trends Biotechnol. 2 : 65-70.
4. Callaham D, Del Tredici P & Torrey JG. 1978. Science 199 : 899-902.
5. Vose PB. 1983. Can. J. Microbiol. 29 : 837-850.

THE PROSPECTS FOR *nif* GENE EXPRESSION IN PLANT CELLS

M. MERRICK, R.A. DIXON, S.B. SHARMA AND J.R. POSTGATE
AFRC UNIT OF NITROGEN FIXATION, UNIVERSITY OF SUSSEX, BRIGHTON, BN1 9RQ, U.K.

INTRODUCTION

Our knowledge of the *Klebsiella pneumoniae* nitrogen fixation (*nif*) genes and the factors controlling their expression has now reached a stage[1] where we can contemplate exploiting this knowledge, together with recent advances in plant molecular biology, to initiate experiments on the transfer of *nif* genes to higher plants.

MANIPULATION OF *nif* GENES

Of the three plant genomes, nuclear, mitochondrial and chloroplast, the chloroplast is the only one which is markedly prokaryotic in nature[2] and consequently this organelle is the location of choice for the *nif* genes. The promoters of the 7 *nif* structural gene operons in *Klebsiella pneumoniae* are all subject to positive control and activation of transcription initiation at these promoters is absolutely dependent on the products of the *nifA* and *ntrA* genes (see Merrick *et al.*, Poster Abstract - this volume).

If the *nifA* and *ntrA* genes could be expressed in a chloroplast and if their products could interact correctly with chloroplast RNA polymerase, it is possible that all the *nif* structural genes could be transcribed and translated without recourse to further manipulation. Such an approach has the attraction of maintaining the inherent regulatory features of the *nif* DNA sequences intact and causing minimal alteration to this finely balanced gene complex.

An appropriate gene package would therefore have the following properties. The *nifA* and *ntrA* genes would be linked to suitable chloroplast promoters. These constructs would in turn be linked to the 15 *nif* structural genes, a chloroplast origin of replication and a selectable chloroplast marker gene. The introduction of such a package into the plant would depend on development of methods for chloroplast transformation. The unicellular alga *Chlamydomonas rheinhardii* is a potential model organism for chloroplast transformation. It has a single large chloroplast per cell and the chloroplast genome has been well characterised. Nuclear transformation has been demonstrated in this organism[3] and selectable chloroplast markers in the form of 'herbicide resistance' mutations in the chloroplast *psbA* gene have recently been described.[4] We have therefore examined chloroplast promoters from *C. rheinhardii* as potential promoters for *nifA* and *ntrA* fusions.

The RUBPcase LS gene (*rbcL*) from maize and wheat can be expressed in *E. coli*, apparently from a chloroplast promoter[5] and we therefore examined *C. rheinhardii* chloroplast promoters for similar activity in *E. coli*. Translational fusions were made between *lacZ* and the promoters and N-terminal coding regions of the *rbcL*, *atpB* and *psbD* genes. Expression of *lacZ* was not observed in *rbcL* and *psbD* fusions but some activity was obtained from *atpB*::*lacZ* fusions. The significance of these observations with respect to transcription and translation initiation signals in *C. rheinhardii* is currently under investigation and further information on the factors controlling gene expression in chloroplasts would clearly be of use in designing future gene fusions.

REFERENCES

1. Dixon, R. 1984. *J. Gen. Microbiol.* *130*, in press.
2. Whitfield, P.R., Bottomley, W. 1983. *Ann. Rev. Plant Physiol.* *34*, 279-310.
3. Rochaix, J-D., van Dillewijn, J. 1982. *Nature*, *296*, 70-72.
4. Erickson, J.M. *et al.* 1984. *Proc. Natl. Acad. Sci.* *81*, 3617-3621.
5. Gatenby, A.A. *et al.* 1981. *Nature*, *291*, 117-121.

MOLECULAR CLONING WITH MITOCHONDRIAL DNA

K.ESSER
Lehrstuhl für Allgemeine Botanik, Ruhr-Universität,
Postfach 102148, D-4630 Bochum, Germany.

The in vitro transfer of DNA was mainly developed by using bacteria as host cells. Accordingly, suitable vector systems for recombinant DNA technology are predominantly of prokaryotic origin. Since gene cloning is becoming an essential component of both fundamental research and applied research, the inclusion of eukaryotes in the spectrum of host organisms becomes necessary. One of the main reasons to explore eukaryotes for such research is that bacteria, when used as hosts for the cloning of eukaryotic DNA, may cause difficulties, such as failure in replication or expression of the foreign DNA, complications when extracting the product, or instability of the transformants. To avoid complications of this kind, it is desirable that the cloning of eukaryotic DNA be restricted to eukaryotic systems only, i.e. eukaryotes providing both the host and the vector.

A new perspective was opened up when a new group of eukaryotic plasmids associated with mt DNA was detected. After the discovery that the filamentous fungus Podospora anserina harbors a plasmid originating from mt DNA, comparable genetic elements were found in some other fungi and also are known to be present in higher plants. Further, it became evident that not only mt plasmids but also segments of mt DNA are suitable for the construction of eukaryotic, autonomously replicating vectors.

There are two sources for the development of vectors in gene cloning from mt DNA: (i) mitochondrial plasmids and (ii) sequences of mt chromosomal DNA which carry a replicon. In the first case, naturally occurring mt plasmids are used; in the second case, a more laborious procedure is required. Since the mt chromosome of each organism carries at least one replicon, in principle mt DNA from any organism may be used to construct a vector. Thus, every organism becomes available for genetic engineering regardless of whether it contains a plasmid or viruses. Until now, both alternative ways of constructing vectors from mt DNA have been provided only in some fungi.

Although the experimental data are far from being comprehensive, they prove that the concept of using mt plasmids or mt DNA as a base to construct autonomously replicating vectors is of general applicability. Naturally, other extrachromosomal genetic traits may also be used in a comparable way for gene cloning. The occurrence of these elements, however, is as yet limited to a few organisms, whereas mt DNA is available from any eukaryotic organism.

REFERENCE

Esser K, Kück U, Stahl U, Tudzynski P. 1983. Cloning vectors of mitochondrial origin for eukaryotes: a new concept in genetic engineering. Curr.Genet. 7, 239-243.

RECOMBIBATION IN WHEAT MITOCHONDRIAL DNA.

F. QUETIER and B. LEJEUNE.

Laboratoire de Biologie Moléculaire Végétale,LA CNRS n° 40, Université Paris-Sud, Orsay. France.

The analysis of a cosmid library of wheat has revealed that i) this genome contains 52 different Sal I fragments whose molecular weights range from 0.4 kb to 29 kb and ii) that a master circular molecule of 430 kb can be predicted (1). Cross-hybridizations systematically achieved between all these individually cloned Sal I fragments have indicated that 26 fragments can hybridize to another or several fragments, thus evidencing the occurence of repeated sequences in the mt DNA. The detailed mapping of all these fragments indicated that these repeated sequences can be classified into 2 types, depending upon whether they contain an internal Sal I site or not; most of them concern sets of 4 fragments wich share a sequence, the length of which being either in the range of several kb (fragments carrying r-RNA genes) or shorter than 0.5 kb.

The comparison of the detailed mapping of each of the 4 Sal I fragments of a set evidenced the occurence of a central common area flanked by differential leftward and rightward stretches. This structure is obviously related to a recombination process as shown in fig 1 for the 4 fragments carrying the 18 S- 5 S r-RNA tandem genes (2).

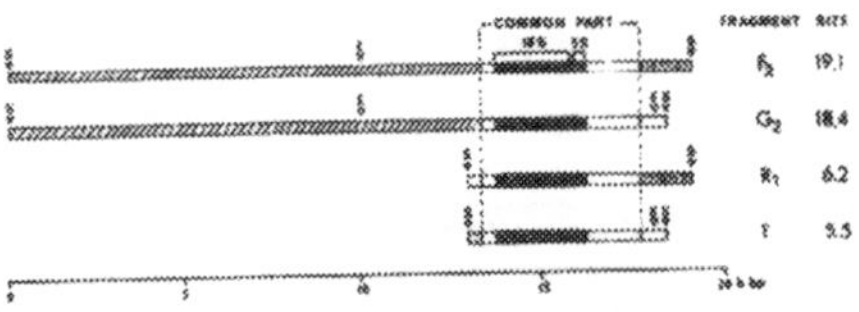

Fig. 1

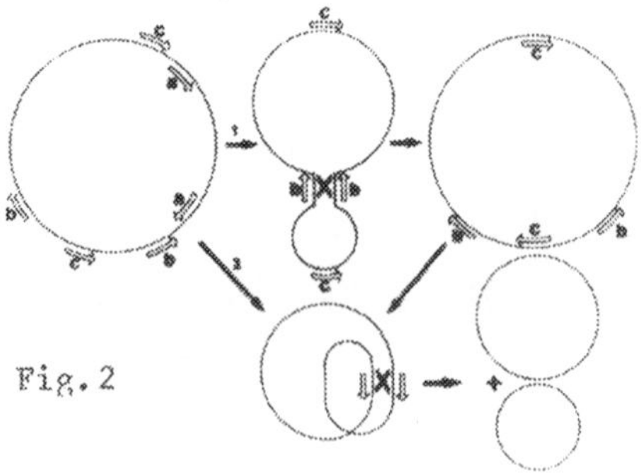

Fig. 2

There are 10 different kinds of such repeated sequences; they have been all located on the master circle of 430 kb . Each kind of repeated sequence is present in 2 copies within this master circle except the repeat habouring the 18S - 5S tandem which occurs 3 times. All the Sal I fragments which are not present in the master circle, but occur in the library, are generated through these recombinational events.

Two kinds of recombinational events can occur (3), depending on the relative orientation of the 2 copies involved in the crossing-over (fig. 2). A pair of direct repeats (a) can give rise to 2 different molecules. A pair of inverted repeats (b) only induces a flip-flop conversion (4)but can thereby convert another pair of inverted repeats (c) into a pair of directs repeats which can now generate 2 diferent molecules.

According to the presence of 10 kinds of repeats, it is clear that many different molecules can be generated through these recombinations, giving rise to the heterogeneous populations of molecules first evidenced from restriction patterns (5) and directly observed by electron microscopy (3). This mechanisms of recombination have been presented here in the case of a circular molecule but it should be outlined that they work with linear molecules as well; this point seems important since the population of mt DNA molecules contains a high proportion of linear molecules (3).

These results raise the question of the replication autonomy of the molecules generated through such recombinations. Restriction patterns are quite reproducible for a given tissue in given physiological conditions, so that unknown mechanisms should exist which maintain the relative stoichiometry of the different molecules roughly constant. The identification of ORI sequence(s) is thus a key to this problem.

References.

1. Quétier, F., Lejeune, B., Falconet, D. and Jubier, M.F. (1984),in Molecular Form and Function of the Plant Genome, Van Vloten-Doting,L.,Groot, G. and Hall, T., Eds. Plenum Press, in press.
2. Falconet, D., Lejeune, B., Quétier, F. and Gray,M.W. (1984), EMBOJ. ,3 ,297-302.
3. Quétier F., Lejeune B., Delorme S., and Falconet D., (1984), Molecular organization and expression of the mitochondrial genome of higher plants. Encyclopedia of Plant Physiology, New Series vol.18, Springer Verlag, in press.
4. Palmer, J.D. and Shields, C.R. (1984) Nature 307, 437-440.
5. Quétier, F. and Vedel F., (1977) Nature 268, 365-268.

POSTERS

REGULATORY FACTORS NECESSARY FOR TRANSCRIPTION OF NITROGEN FIXATION (nif) GENES: IMPLICATIONS FOR GENETIC MANIPULATION

M. MERRICK, M. BUCK, R.A. DIXON, M. DRUMMOND, T. HAWKES AND J.R. POSTGATE
AFRC UNIT OF NITROGEN FIXATION, UNIVERSITY OF SUSSEX, BRIGHTON, BN1 9RQ, U.K.

INTRODUCTION

The free-living nitrogen-fixing bacterium Klebsiella pneumoniae has provided a model system for analysis of the organisation and regulation of the nif genes. These studies are an essential prerequisite for any subsequent manipulation of the nif gene cluster in order to develop a 'gene package' suitable for introduction into a plant cell. Genes specifically required for the synthesis and activity of nitrogenase form a complex regulon of 17 nif genes organised into 8 transcriptional units. Nitrogenase synthesis is stringently controlled in response to extracellular levels of fixed nitrogen or oxygen, and the products of five genes are required to regulate nif transcription. Three of these genes ntrA,B,C are required for control of general nitrogen assimilation in the cell and are unlinked to the nif regulon. Two other regulatory genes, nifLA, are located in the nif gene cluster.

Regulation is a two-tier cascade process whereby activation of nifLA transcription in nitrogen-limiting conditions requires the ntrC and ntrA gene products. The nifA and ntrA products, in turn, are required for activation of the other nif promoters.[1,2] The requirement for ntrC product can be circumvented by linking nifA to a constitutive promoter although, as predicted, in such constructs nifA activation is still ntrA-dependent. In the presence of repressive levels of extracellular oxygen or fixed nitrogen the nifL product is thought to inactivate nifA product and prevent further nif transcription.[3] Likewise, ntrB product responds to fixed nitrogen and inactivates the ntrC product thereby preventing nifLA transcription.

Comparative sequence analysis of nif promoters reveal striking sequence homology between them but little homology with the consensus sequence for other prokaryotic promoters[4,5] (Table 1). Current studies suggest that the ntrA product may act as a specific sigma factor to modify the transcription initiation specificity of RNA polymerase at these atypical promoters.

TABLE 1. Comparison of ntrA-dependent and ntrA-independent promoters.

	-35	-10
ntrA-independent (E. coli consensus)	TTGaca	TAtaaT
	-24	-12
ntrA-dependent (nif promoters)	ctGGca	ttGCa

The model we have proposed for activation of nif transcription has important implications for obtaining expression of the Klebsiella nif gene cluster in other 'prokaryotic-like' environments, e.g. the plant chloroplast. We suggest that the nif operons might be expressed without major modification if the nifA and ntrA genes were fused to chloroplast promoters so as to synthesise those necessary regulatory proteins in the plastid. The choice of plastid promoter could offer a degree of spatial or temporal regulation of nif within the plant. The success of such an approach would depend on the compatibility of the nifA and ntrA products with chloroplast RNA polymerase which although similar to the bacterial enzyme may be somewhat more complex.

REFERENCES

1. Merrick, M. 1983. EMBO J. 2, 39-44.
2. Ow, D., Ausubel, F.M. 1983. Nature, 301, 307-313.
3. Merrick, M. et al. 1982. Mol. Gen. Genet. 185, 75-81.
4. Drummond, M. et al. 1983. Nature, 301, 302-307.
5. Beynon, J. et al. 1983. Cell, 34, 665-671.

BIOCHEMICAL AND GENETIC ANALYSIS OF NITROGEN FIXATION IN SESBANIA RHIZOBIUM

C. ELMERICH, F. NOREL, A. KUSH, P. DENEFLE, N. CHARPIN, A. PAQUELIN AND J.-P. AUBERT.

INTRODUCTION

The fast growing *Rhizobium* strain ORS571, isolated from stem nodules of the tropical legume *Sesbania rostrata* was shown to form nodules both on the stem and on the roots of the host plant, and to be able to grow in the free living state at the expense of N_2 as sole nitrogen source (1).

1. PROPERTIES OF THE NITROGENASE

The nitrogenase was purified from N_2-grown cells. The enzyme is very similar to previously characterized nitrogenases. Addition of NH_4^+ to cultures of N_2-growing cells resulted in an immediate and practically total inhibition of nitrogenase activity, either in whole cells or in crude extracts. Reactivation of nitrogenase in crude extracts was dependent on an O_2-labile factor present only in N_2-grown cells. Purification of the activating factor is in progress.

2. HOMOLOGY WITH *K. PNEUMONIAE NIF* SEQUENCES

DNA homology with *Klebsiella pneumoniae nif* probes appears to be extended to more genes than in other *Rhizobia*. Homology in total DNA was detected by Southern hybridization, with *nifHDK* (pSA30), *nifNE* (pGR113) and *nifJ* (pPC880). DNA fragments carrying some of the homologous sequences were cloned. Plasmid pRS2 contains a 13 kb *Bam*HI fragment that carries homology with *nifHDK* (2) and with *nifNE*. On this fragment, homology with *nifNE* was located on a 2.3 kb *Bam*HI-*Bgl*II fragment, whereas homology with *nifHDK* was located on the adjacent *Bgl*II-*Sal*I fragment of 4 kb. Precise localization of the *nifHDK* genes was established using specific *nifH*, *nifD* and *nifK* probes. Homology with *nifH* was found on a 0.6 kb *Xho*I fragment. *NifD* and *nifK* were located on a *Bgl*II-*Xho*I fragment between *nifNE* and *nifH*. Thus, in strain ORS571 the *nifHDKNE* genes are likely to be clustered in the same order as in *K. pneumoniae*.

Moreover, hybridization results are consistent with the existence of a single copy of *nifK* and *nifD* whereas *nifH* and sequences homologous to *nifJ* are reiterated in ORS571 total DNA. Another copy of *nifH* unlinked to that cloned in pRS2 was identified, and three *BamH1* fragments containing DNA sequences homologous to *nifJ* were cloned.

3. ISOLATION OF NIF^- MUTANTS BY TRANSPOSON SITE DIRECTED MUTAGENESIS

Transposition of Tn*5*, in plasmid pRS8, was selected in *E. coli*. This plasmid carries the *nifHDKNE* cluster of strain ORS571 cloned in pRK290. Tn*5* insertions were mapped by restriction analysis and localized in *nifD*, *nifK* and *nifNE* genes. A specific deletion of *nifH* was obtained by *in vitro* substitution of the 0.6 kb *XhoI* fragment by a 2.3 kb *XhoI* fragment carrying the Km^r gene of Tn*5*. The mutagenized plasmids were transferred in strain ORS571. Tn*5* insertions in the host genome were obtained by marker exchange after surinfection with pPH1, a plasmid incompatible with pRK290. These experiments were performed only for insertions in *nifD* and *nifK*. The resulting strains were Nif^-. Product analysis of the mutants and further experiments to obtain deletion of *nifH* and insertions in *nifNE* are in progress.

REFERENCES

1. Dreyfus BL, Elmerich C and Dommergues YR. 1983. Free-living *Rhizobium* strain able to grow on N_2 as the sole nitrogen source. Appl. Environm. Microbiol. 45 : 711-713.

2. Elmerich C, Dreyfus BL, Reysset G and Aubert J.-P. 1982. Genetic analysis of nitrogen fixation in a tropical fast-growing *Rhizobium*. EMBO J. 1 : 499-503.

THE pSym OF RHIZOBIUM MELILOTI 2011 : PHYSICAL MAPPING OF A 290 kb BLOCK OF pSym AND LOCALIZATION OF SYMBIOTIC GENES OUTSIDE THE nod-nif REGION.

T. HUGUET, M. GHERARDI, B. TERZAGHI*, D. KAHN, M. DAVID, O. DOMERGUE, E. TERZAGHI**, A.M. GARNERONE, P. BOISTARD and J. BATUT.
Laboratoire de Biologie Moléculaire des Relations Plantes-Microorganismes, CNRS-INRA, B.P.12, 31320 Castanet-Tolosan, France.

R. meliloti fixes nitrogen in symbiosis with its host plant alfalfa. In R. meliloti genes controlling the nodule formation (nod genes) and fix genes controlling the nitrogen fixation process (such as the nifHDK nitrogenase structural genes) are closely linked on a large size megaplasmid (>700 kb) called pSym. (1,2,3)

We previously constructed a plasmid, pGMI42, which contains a 290 kb fragment of R. meliloti pSym cloned in vivo in the broad host range plasmid RP4 (4). Using a site-directed in vivo cloning procedure involving homologous recombination we showed that the cloned fragment in pGMI42 is colinear with the corresponding region of pSym. Physical mapping of pGMI42 was done by cloning HindIII partial digests into pBR322. The 290 kb insert starts from 5 kb to the right of nifHDK and extends to the left of nifHDK including genes controlling host-specific nodulation (5).

We demonstrated that, in addition to the nod-nif region, other regions of pSym are actively transcribed in bacteroids. Tn5 site-directed mutagenesis was performed to study one of these transcribed regions located 200 kb from nifHDK. Four mutants defective in nitrogen fixation were characterized. The linkage between the Tn5 insertion and the Fix^- phenotype was demonstrated in each case by complementation with plasmids carrying the wild-type sequence.

REFERENCES
(1) Banfalvi et al. 1981. Mol. Gen. Genet. 184 : 334-339.
(2) Rosenberg et al. 1981. Mol. Gen. Genet. 184 : 326-333.
(3) Corbin et al. 1983. Proc. Natl. Acad. Sci. 80 : 3005-3009.
(4) Julliot et al. 1984. Mol. Gen. Genet. 193 : 17-26.
(5) Truchet et al. 1984. J. Bact. 157 : 134-142.

*Present address : Plant Physiology Division, DSIR, Private Bag, Palmerston North, New-Zealand.
**Present address : Massey University, Palmerston North, New-Zealand.

GENETIC ANALYSIS OF THE ADENYL CYCLASE (*cya*) REGION OF *RHIZOBIUM MELILOTI*

F. O'GARA, R. LATHIGRA, M. O'REGAN, B. KIELY AND A. McGETTRICK
MICROBIOLOGY DEPARTMENT,
UNIVERSITY COLLEGE,
CORK,
IRELAND.

INTRODUCTION

During the complex interaction in the *Rhizobium*-legume symbiosis the pattern of gene expression in both the bacterial and plant partners undergoes a differential switch to facilitate the provision of cellular and metabolic activities necessary for symbiosis. The nature of the signals involved in mediating this process of differential gene expression is not understood. Cyclic AMP is well recognised as an important effector molecule in controlling gene expression in *Escherichia coli* and some other Gram negative bacteria (1, 2). Although the nucleotide is detectable in *Rhizobium* strains its involvement in symbiosis is presently unclear. The recent isolation of a DNA fragment encoding adenyl cyclase sequences from *R. meliloti* (3) provides the basis to elucidate fundamental principles in this area.

RESULTS AND DISCUSSION

Previous data (3) established that the *R. meliloti* cya gene was located on a 5.47 Kb *Bgl*II fragment cloned in pRK290, a low copy number broad host range vector. This *cya* fragment weakly complemented *E. coli* Δ*cya* mutations. A restriction map of the fragment designated 2046 is shown in Fig. I. Two large subfragments (A and B) are produced in double digests of 2046 with *Bgl*II and *Bam*HI restriction endonucleases. These fragments can be conveniently cloned into the *Bam*HI site of pBR322. The *cya* gene was localized to the 2.6 Kb *Bgl*II-*Bam*HI DNA region (fragment A) by subcloning and insertional mutagenesis with transposons Tn5 and Mudl (Ap^r *lac*). 'Invivo' transcription-translation experiments in *E. coli* 'maxicells' showed that the *R. meliloti cya* gene product is 64,000 daltons which is significantly smaller than the corresponding protein from enteric bacteria. Fragment B of 2046 which does not complement

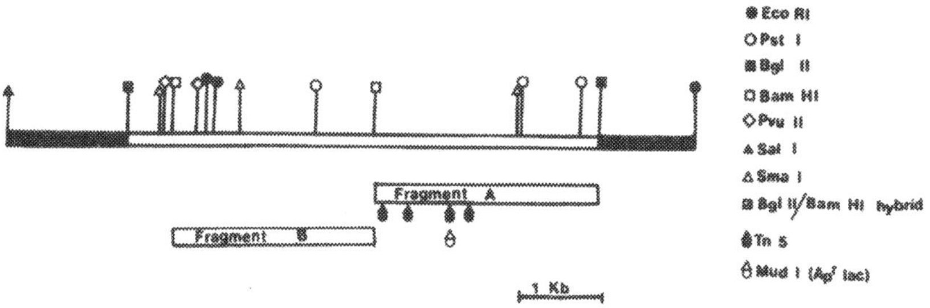

FIGURE 1. Restriction map of R. meliloti cya fragment 2046. Position of Tn5 and Mud-1 on fragment A were determined by restriction enzyme analysis. ■ = pRK290 vector.

Δcya mutations directs the synthesis of a 43,000 dalton protein the function of which remains unknown. Fragments A and B of 2046 were cloned into the transcription-translation fusion vector pMC1403 in both orientations yielding a series of blue and white clones on X-gal plates. Both fragments contain promoters and translation initiation sites which are recognised in an E. coli background. The direction of transcription of the two promoters is similar. On fragment A transcription proceeds from the BglII to the BamHI site while on fragment B transcription is from the BamHI towards the EcoR1 sites. Results from DNA hybridization experiments indicated that although both the A and B fragments of 2046 are highly conserved in R. meliloti strains no evidence of significant sequence homology could be detected in other rhizobial species or E. coli under the conditions employed. This may indicate that there are differences in the adenyl cyclase proteins among Rhizobium species.

REFERENCES

1. Botsford JL, 1981. Cyclic nucleotides in prokaryotes. Microbiol Rev 45: 620-642.
2. Ullman A, Danchin A, 1983. Role of cyclic AMP in bacteria. Adv cyclic nucleotide Res, V15. Raven Press, New York.
3. Kiely B, O'Gara F, 1983. Cyclic 3'5' adenosine monophosphate synthesis in Rhizobium : Identification of a cloned sequence from Rhizobium meliloti coding for adenyl cyclase. Mol Gen Genet 192: 230-234.

ANALYSIS OF THE NODULATION GENES OF RHIZOBIUM LEGUMINOSARUM

J. A. DOWNIE, C. D. KNIGHT, L. ROSSEN AND A. W. B. JOHNSTON

1. INTRODUCTION

The ability of strains of *Rhizobium leguminosarum* to nodulate peas and to fix nitrogen is determined by the presence of large native symbiotic plasmids. In one such plasmid, pRL1JI, the symbiotic genes are within a 60 kb region with the nodulation genes located between two groups of genes required for nitrogen fixation. By cloning the nodulation DNA in wide host-range plasmid vectors and transferring the cloned DNA to other species of *Rhizobium* with different host-ranges or to strains that were cured of their native symbiotic plasmids it was shown that less than 10 kb was required for the determination of nodulation ability. In this presentation, we report on the detailed physical and genetic analyses that have been conducted on this nodulation region of pRL1JI.

2. RESULTS AND DISCUSSION

The cloned nodulation DNA was extensively mutagenised with Tn5 and these mutations were introduced by homogenotisation into the symbiotic plasmid pRL1JI. The various mutations were analysed (a) by mapping their locations and (b) by observation of their effects on the nodulation process; this was done both by microscopic examination of the inoculated roots and by enumerating the nodules that were induced. On these bases five different classes of nodulation deficient mutations were identified. Two of the classes failed to nodulate; one of these failed to induce any root hair formation whereas the other induced slight root hair curling but only after a delay. The other three classes did

The nodulation DNA was used as a template in an *in vitro* transcription/translation system and the polypeptides formed were compared with those produced when the nodulation DNA included various of the *nod::Tn5* alleles. By such means it was shown that one of the classes of mutant alleles that stopped nodulation was defective in the synthesis of a 46 K polypeptide whereas the other class of *nod::Tn5* alleles which prevented nodulation failed to make a 34K polypeptide. By sequencing the DNA of this region, open reading frames that corresponded to the genes that were identified by genetic methods were observed.

Although the nitrogen-fixing legume nodule is a complex organ, it is apparent that relatively few *Rhizobium* genes are required for this bacterium to infect the host plant and to induce the formation of nodules. Thus it has proved possible to analyse in detail the nodulation genes that confer on *Rhizobium* its unique ability to participate in this very important symbiotic interaction.

HOST RANGE GENES IN *RHIZOBIUM LEGUMINOSARUM*

R. GÖTZ, I. EVANS, J. A. DOWNIE, AND A. W. B. JOHNSTON

1. INTRODUCTION

Some primitive pea lines, e.g. cultivar Afghanistan are not nodulated by "normal" strains of *Rhizobium leguminosarum* (from Western Europe), but strain TOM, from Turkey, does nodulate both primitive and cultivated pea lines. The genes for nodulation and host plant specificity of strain TOM have been cloned on plasmid pIJ 1095 which contains about 30kb of DNA from the symbiotic plasmid of strain TOM. When mobilised into a *Rhizobium* strain that had been cured of its symbiotic plasmid, pIJ 1095 conferred upon its recipient the ability to nodulate both cv. Afghanistan and cultivated lines of peas. Derivatives of a "normal" *Rhizobium leguminosarum* strain containing the cloned *nod* region from strain TOM (pIJ 1095) gained the ability to form effective nodules on cv. Afghanistan (1).

2. RESULTS

Several nodulation deficient derivatives of plasmid pIJ1095 containing Tn*5* in the cloned *nod* genes were isolated and grouped into three phenotypic classes (1). One mutant, *nod*-13::Tn*5*, was specifically affected in the host-range determinant, because transconjugants of a "normal" *Rhizobium leguminosarum* strain (e.g. a strain with the symbiotic plasmid pRL1JI) carrying the *nod*-13::Tn*5* mutant plasmid could not nodulate cv. Afghanistan. The *nod*-13::Tn*5* mutant allele was homogenotised from pIJ 1095 into the TOM symbiotic plasmid (pRL5JI) by marker exchange and the resultant strain analysed for its nodulation phenotype. On cv. Afghanistan nodulation was delayed by about three days and the number of nodules formed was reduced by about 70%. However on the commercial pea cultivar (Wisconsin Perfection) nodulation was delayed only by about one day and the number of nodules was reduced by only 20%. The induction of root hair curling by the strain carrying the nod-13::Tn5 allele was compared with the wild-type strain and the other classes of nod mutants (classes I and II) which had been identified. Whereas mutants of classes I and II were either deficient or abnormal in their ability to induce root hair curling, the nod-13::Tn5 mutant strain was normal.

From this analysis it appeared that the two functions, root hair curling and host-range, could be separated. Two overlapping subcloned wild-type fragments spanning the region around the nod-13::Tn5 allele were shown to confer upon a "normal" Rhizobium leguminosarum strain the ability to nodulate cv. Afghanistan. Because neither subcloned fragment could induce nodules when present in a strain without a symbiotic plasmid, it was clear that they carry only some nod genes and these determine host plant specificity. By deletion mapping, the cv. Afghanistan pea host range determinant has been localised within a region of 2.8kb of DNA. Using DNA hybridisation, and by comparing restriction enzyme patterns, it has been established that there is homology between this host range region and part of the nod region of a strain which is unable to nodulate cv. Afghanistan. The extent and significance of this homology with regard to different host plant specificities is currently under investigation.

REFERENCES

1. Hombrecher G, Götz R, Dibb NJ, Downie JA, Johnston AWB, Brewin NJ. 1984. Cloning and mutagenesis of nodulation genes from Rhizobium leguminosarum TOM, a strain with extended host range. Molecular and General Genetics 194:293-298.

MOLECULAR BIOLOGY OF COMPETITION AMONGST *RHIZOBIUM* FOR NODULATION OF *PISUM SATIVUM* CV. AFGHANISTAN

D.N. DOWLING, U. SAMREY, and W.J. BROUGHTON
Max-Planck-Institut für Züchtungsforschung, D-5000 Köln 30, FRG

1. INTRODUCTION

Nodulation of the primitive pea *Pisum sativum* cv. Afghanistan by *R. leguminosarum* strain TOM can be blocked by a European *R. leguminosarum* strain PF_2, which does not form nodules on cv. Afghanistan.

The nature of this blocking event is unknown; it has been shown that PF_2 does not produce compounds inhibitory to TOM (1) but PF_2 does accumulates on pea roots faster than TOM (2). The presence of living PF_2 cells is necessary for the blocking process for although dead cells can bind to the root they no longer block nodulation by TOM.

Thus this nodulation blocking system is a suitable model to study the bacterial component of competition. The localization and characterization of the gene(s) involved in this process would be of great interest in defining the mechanisms of competition between *Rhizobium* strains.

2. RESULTS AND DISCUSSION

As plasmids play a role in the symbiosis in *R. leguminosarum* we decided to first investigate if nodulation blocking was plasmid encoded in PF_2. PF_2 was found to have 4 large plasmids and one of these hybridized to *nif* and *nod* probes. The plasmids present in PF_2 were labelled with the mobilizing transposon Tn5-Mob (3) and mobilized by RP4.4 to the plasmid free *Rhizobium* NZP4010. In this way recipient strains were constructed with different combinations of PF_2 plasmids. These are being screened for their ability to block nodulation by TOM. Preliminary data suggests that one of these plasmids (which hybridized with *nif* and *nod*) may play a role in blocking.

In order to clone the gene(s) involved in nodulation blocking, a cosmid bank of total PF_2 DNA is under construction in the wide host range cosmid vector pMMB33 (4). The bank will be mobilized into suitable *R. leguminosarum* and *R. trifolii* recipients and screened for the ability to block TOM from nodulating Afghanistan peas.

3. REFERENCES

1. Broughton WJ, Samrey U, and Bohlool BB. 1982. Can. J. Microbiol. 28, p 162-168
2. Broughton WJ, van Egeraat AWSM, and Lie TA. 1980. Can. J. Microbiol. 26, p 562-565
3. Simon R, Priefer U, and Pühler A. 1983. In: Molecular Genetics of the Bacteria-Plant Interaction (ed. A. Pühler) p 98-106, Springer-Verlag, Berlin-Heidelberg
4. Frey J, Bagdasarian M, Feiss D, Franklin CH, and Deshusses J. 1983. Gene 24, p 299-308

GENETIC CHARACTERIZATION OF THE RHIZOBIUM/VICIA FABA SYMBIOSIS

U.B. PRIEFER, S. MOHAPATRA, J. KUHSE, AND A. PÜHLER
Lehrstuhl für Genetik, Fakultät für Biologie, Universität Bielefeld, Postfach 8640, D-4800 Bielefeld

1. INTRODUCTION

The leguminous plant Vicia faba is becoming increasingly interesting and important for agriculture in European countries. The project, started in our laboratory in April '84, is focussed on the genetic analysis of the interaction between V.faba and Rhizobium. For this purpose, bacterial as well as plant components are being investigated. Our preliminary data and results on both aspects will be presented.

2. ISOLATION AND ANALYSIS OF NODULE BACTERIA

We have isolated a number of Rhizobium strains from nodules of V.faba. All isolates were able to nodulate several different cultivars of V.faba and P.sativum. Some of these strains have been tested by hybridization for the presence of genes homologous to R.meliloti nif and fix genes (see accompaning poster by S.Manian and A.Pühler). The plasmid content of the strains was determined using a modified Eckhardt lysis technique (1,2). The plasmid profile varied considerably and the number of plasmids ranged from three to eight. We also screened some of the strains for the presence of transmissible plasmids by carrying out random Tn5 mutagenesis (3), testing for transfer of Nm resistance in batch crosses with various recipients. In one strain (VF39) two selftransmissible plasmids were detected which have approximate molecular weights of 100 and 160 Md. The smaller of these two plasmids appears to carry genes involved in nodulation.

3. IDENTIFICATION OF NODULE SPECIFIC PROTEINS

Development of functional root nodules involves specific and amplified synthesis of several polypeptides by both plant and bacterium. Detection and enumeration of these by immunoblotting technique constituted the first step in molecular analysis of plant genes involved in symbiosis (4,5).

Antisera were raised against nodule proteins and used to analyze the polypeptides from V.faba nodules. Most of these comigrated with either root or bacterial polypeptides. At least twelve of the polypeptides (ranging from 14 K to 130 K) appeared to be plant encoded. Of these, five, including leghemoglobin, were smaller than 20 K. Whether or not these polypeptides are plant encoded is being confirmed by in vitro translation of poly A^+ RNA extracted from root and nodule tissue. Alfalfa nodule specific antisera recognized some polypeptides from V.faba nodules, indicating that these nodule specific polypeptides may be common to several legumes.

4. MOLECULAR ANALYSIS OF A C-DNA CLONE CODING FOR LEGHEMOGLOBIN

Leghemoglobins are the most prominent proteins in root nodules induced by Rhizobium species. We have begun studying leghemoglobin genes of V.faba. Poly A^+ RNA isolated from polysomes of V.faba root nodules was used to construct a c-DNA library in the vector pBR322. A total of 1600 c-DNA clones was screened for leghemoglobin sequences by hybridization with pPSLB101, a c-DNA clone carrying the pea Lb gene (kindly provided by T.Bisseling). One clone, pCK2-77, was further characterized by northern blot hybridization, sequencing, and hybrid released translation experiments. We also sequenced a second c-DNA clone, pCK8-61, identified by hybridization with pCK2-77. The two clones share base pair homology ranging from 30 % to 75 %. However, the amino acid sequence deduced from pCK8-61 differed markedly from that of the Lb clone pCK2-77.

We wish to acknowledge one of our students, Mrs. K. Bruksch, who did the analysis and characterization of the V.faba isolates.

REFERENCES

(1) T. Eckhardt (1978). Plasmid 1:584
(2) M.F. Hynes et al. (1984). Plasmid, submitted
(3) R. Simon et al. (1982). Biotechnology 1:784
(4) T. Biseling et al. (1984). In: Advances in Nitrogen Fixation Research (C.Veeger, W.E.Newton, eds.), Nijhoff/Junk Publ., The Hague, 579
(5) S.S. Mohapatra and A. Pühler (1984). In: Abst. 2nd Int. Meeting on Molecular Genetics of the Bacteria-Plant Interaction, Cornell University, Ithaca, 84

IDENTIFICATION AND LOCALISATION OF FIX GENES OF THE VICIA FABA MICROSYMBIONT

S.S. MANIAN/A. PÜHLER
Lehrstuhl für Genetik, Fakultät für Biologie, Universität Bielefeld, Postfach 8640, D-4800 Bielefeld, FRG

1 INTRODUCTION

Recently a programme aimed at improving the efficiency of the Vicia faba-Rhizobium symbiosis has been initiated in our laboratory. As part of this programme a genetic analysis of the Vicia faba microsymbiont is being carried out. In Rhizobium meliloti the main nif/fix region has been shown to contain at least 8 genes (1). The order and direction of transcription of these genes is shown in Fig. 1. Information from sequencing data suggest the possibility of another fix gene adjacent to fixD (1). Several of these fix genes have now been identified by DNA hybridisation studies in Vicia faba microsymbionts.

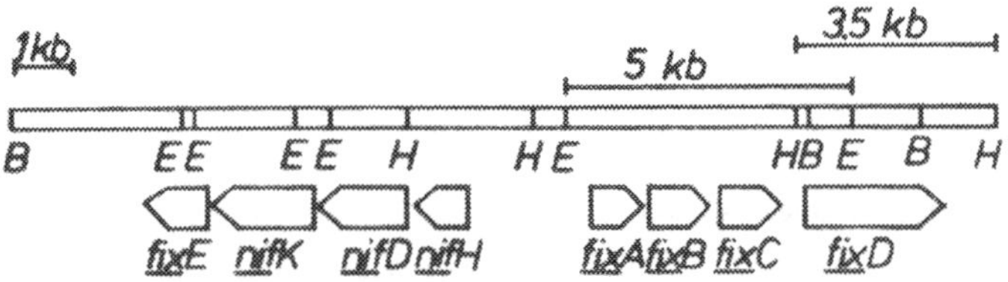

Figure 1: The main nif/fix region of R.meliloti. B: BamHI, E: EcoRI, H: HindIII.

2. MATERIALS AND METHODS

All the Vicia faba microsymbiont strains used in this study are wild type strains isolated in Bielefeld (see Priefer et al., this volume). Plasmid pRmR3 contains a 5.0 kb EcoRI fragment carrying the R.meliloti fixA, fixB and fixC genes (2). Plasmid pRmW54 contains a 3.5 kb HindIII fragment carrying the R.meliloti fixD gene (1).

Isolation of total and plasmid DNA, nick translation and hybridisation to Southern blots were all carried out according to standard published methods (3).

3. RESULTS AND DISCUSSION

Preliminary results indicated that while the *R.meliloti* *fix*A, *fix*B and *fix*C region showed homology to DNA fragments from all the *Vicia faba* microsymbionts studied the *R.meliloti* *fix*D region showed homology to DNA fragments from only some of the strains. In those strains which hybridised to both DNA probes the sizes of the hybridising fragments were generally similar for a given restriction enzyme. Data will be presented on the position of these *fix* genes relative to other symbiotic genes such as the *nod* and *nif* genes. Further work will involve a) the cloning of these *fix* genes, b) the construction of a detailed restriction map of the region and c) site-directed mutagenesis of the region coupled with an examination of the symbiotic performance of the resulting mutants.

4. CONCLUSION

Data are presented on the identification, by hybridisation studies, of DNA fragments of the *Vicia faba* microsymbionts which show homology to the *R.meliloti* *fix*A, *fix*B, *fix*C and *fix*D region. Localisation of the individual *fix* genes within this region requires studies with more specific DNA probes.

REFERENCES

1. Pühler A, Aguilar OM, Kapp D, Reiländer H, Weber G. 1984. Mapping and regulation of nitrogen fixation genes of *Rhizobium meliloti*. In: Proceedings of the 14th International Symposium on the Chemistry of Natural Products, Poznan, Poland.
2. Ruvkun GB, Sundaresan V, Ausubel FM. 1982. Directed transposon Tn5 mutagenesis and complementation analysis of *Rhizobium meliloti* symbiotic nitrogen fixation genes. Cell 29:551-559.
3. Pühler A, Timmis KN. 1984. Advanced Molecular Genetics, Springer-Verlag, Berlin Heidelberg New York.

SCREENING METHOD FOR THE DETECTION OF UPTAKE HYDROGENASE ACTIVITY OF *RHIZOBIUM LEGUMINOSARUM* BACTEROIDS

TICHY, HANS-VOLKER AND WOLFGANG LOTZ

A method has been developed for the screening of *R. leguminosarum* strains or mutants for uptake hydrogenase (*hup*) activity. H_2-dependent methylene blue (MB) reduction has been used to detect hydrogen oxidation. Since *R. leguminosarum* strains do not express uptake hydrogenase genes in the free living state, isolated bacteroids have been used for the assay (D. Arp, pers. comm. 1982). Inhibitors of dehydrogenases (jodoacetic and malonic acids) have been used to reduce endogenous substrate-related MB reduction (1). A simple device has been constructed to sparge bacteroid suspensions (containing MB and inhibitors) with argon (to remove oxygen) and hydrogen. Six strains can be tested at the same time including controls without hydrogen. Ten strains with known Hup phenotypes (determined with the ^3H-uptake method by Louise M. Nelson, pers. comm. 1982) gave the expected results. Several *R. leguminosarum* strains which have been shown to carry *hup*-specific sequences on the *nif* plasmid (2) and some which do not, showed the corresponding Hup phenotype (positive or negative MB reduction).

REFERENCES

1. Haugland, R.A. et al. 1983. Appl. Environ. Microbiol. 45, 892.
2. Seifert, B. et al. 1984. In Advances in Nitrogen Fixation Research (Veeger, C. and W.E. Newton, eds.) M. Nijhoff/Dr. W. Junk, The Hague and Pudoc, Wageningen, The Netherlands, p 721.

CONSTRUCTION OF A RHIZOBIUM LEGUMINOSARUM COSMID GENE BANK AND IDENTIFICATI OF GENES FOR HYDROGEN UPTAKE (HUP)

TICHY, HANS-VOLKER, PETRA RICHTER AND WOLFGANG LOTZ

The R. leguminosarum strain B10 has been isolated from pea root nodules (1) It nodulates pea (Pisum sativum), faba bean (Vicia faba) and Phaseolus bean (Phaseolus vulgaris nânus) and fixes nitrogen in symbiosis with pea. In addition, this strain expresses an uptake hydrogenase activity in pea root nodules (Louise M. Nelson, pers. comm. 1982).

In order to study the molecular genetics of hydrogen uptake in R. leguminosarum, a gene bank of the total DNA of strain B10 has been set up. A partial BamHI digest of Rhizobium DNA has been cloned in the wide host range cosmid vector pMMB34 (2). Both vector and insert DNA have been dephosphorylated according to Ish-Horowicz and Burke (3) in order to avoid cloning artifacts as well as the formation of vector multimers. Sixhundred of the 1000 clones of the gene bank have been screened by dot blot hybridization for the presence of hup genes. A HindIII subclone of the plasmid pHU1 (4) containing R. japonicum hup genes has been used as a probe We have shown previously that the nif plasmids of several Hup$^+$ R. leguminosarum strains (including strain B10) carry sequences homologous to pHU1 whereas Hup$^-$ strains did not (5). The 42 Kbp large cosmid pR1B505 hybridized strongly with the pHU1 subclone used as a probe as well as to other subclones containing EcoRI fragments of pHU1. In addition, the insert DNA of pR1B505 hybridized to the nif plasmid of R. leguminosarum B10, as expected. A restriction enzyme cleavage map of pR1B505 has been constructed for BamHI, HindIII, EcoRI and SstI by partial digestion (BamHI), subcloning of plasmid DNA partially digested with HindIII (HindIII, EcoRI and SstI) and Southern blot hybridizations.In order to demonstrate the existence of functional hup genes on pR1B505 we are now constructing R. leguminosarum mutants by site directed mutagenesis using the cloned hup-specific DNA.

REFERENCES

1. Tichy, H.V. and W. Lotz 1981. FEMS Microbiol. Lett. 10, 203.
2. Frey, J. et al. 1983. Gene 24, 299.
3. Ish-Horowicz, D. and J.F. Burke 1981. Nucleic Acid Res. 9, 2989.
4. Cantrell, M.A. et al. 1983. Proc. Natl. Acad. Sci. U.S.A. 80, 181.
5. Seifert, B. et al. 1984. In Advances in Nitrogen Fixation Research (Veeger, C. and W.E. Newton, eds.) M. Nijhoff/Dr. W. Junk, The Hague and Pudoc, Wageningen, The Netherlands, p. 721.

LEGUME ROOT HAIR PROTEINS

D. WERNER and K.-P. KUHLMANN
FACHBEREICH BIOLOGIE, PHILIPPS-UNIVERSITÄT MARBURG
KARL v. FRISCHSTR., D-355 MARBURG, FEDERAL REPUBLIC OF GERMANY

1. INTRODUCTION

The early events of Rhizobium-legume infection and nodule formation are still mainly descriptive: Recognition, attachment, root hair curling, infection thread formation, root hair nucleus migration, abortion and lysis of many symbionts in the root hair. Infections of other parts of roots in some legumes as Arachis spec. modifies this sequence. Three Rhizobium genes (A, B, C) for nodulation have been characterized (1), however the functions of the proteins produced are still unknown. Progress in understanding root hair reactions to Rhizobium gene products involves complex areas of cell biology as cell wall dissolution and synthesis and cell differentiation (2). Lysis of bacteroids close to the host cell nucleus important during root hair infection can also be studied with Rhizobium mutants in later stages of nodule development (3). Lysis of the peribacteroid membrane can be correlated with another specific plant reaction, the accumulation of the phytoalexin glyceollin I (4).

2. RESULTS AND DISCUSSION

Using standard methods for sterile seed germination, isolated root hairs and total primary roots of Pisum sativum cv Rosa Krone were extracted for total soluble proteins and proteins solubilized after sonification of pellets. Results, as documented in Fig. 1, can be summarized as follows: On SDS-gels, a protein with a molecular weight of 80 000 D is significantly enriched in root hairs (lane 4 and 5) compared to total roots (lane 2 and 3). A protein with a MW of 56 000 D is present in root hairs and roots in about equal concentrations, as estimated from gel staining. A large proportion of proteins or protein subunits from both root parts has a molecular weight of 12 500 D or less (Fig. 1). On other gels, documented on the poster, the protein of 80 000 D from pea root hairs seems to be also somewhat enriched compared to the root hairs from _Triticum aestivum_ cv. Kolibri, the 56 000 D protein is more enriched whereas a major protein with a molecular weight of 61 000 D from wheat root hairs is almost completely absent in pea root hairs. Preliminary results with HPLC-FPLC methods confirm the different protein pattern in root hairs and total roots of legumes.

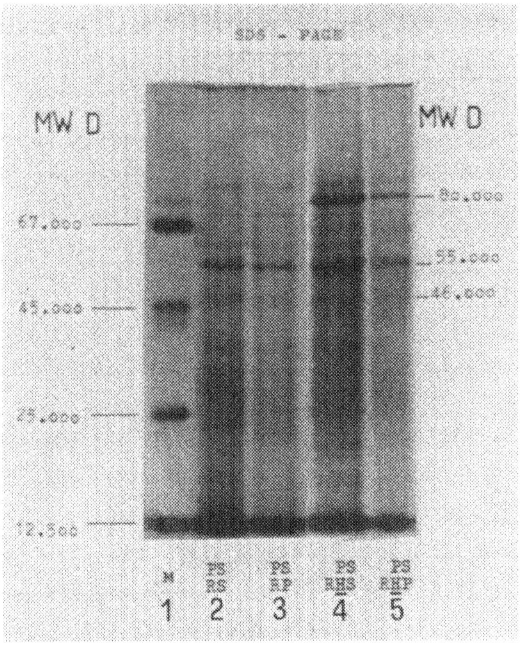

FIGURE 1. SDS gel electrophoresis of protein extracts from _Pisum sativum_ cv. Rosa Krone. Lane 1: marker proteins (67 000 to 12 500 D); Lane 2: total root proteins, soluble. Lane 3: total root proteins, from pellets solubilized. Lane 4: root hair proteins, soluble. Lane 5: root hair proteins, solubilized from pellets. Lane 2-4: 120 to 150 µg protein. Lane 5: 75 µg protein.

REFERENCES

1. Kondorosi A, Banfalvi Z, Broughton WJ, Kondorosi E, Pankhurst CE, Rhandawa GS. 1983. In: Molecular genetics of bacteria plant interaction (A Pühler ed.) Springer.
2. Long S. 1984. In: Plant microbe interactions (E Nester, T Kosuge eds.) Macmillan Press.
3. Werner D, Mörschel E, Kort R, Mellor RB, Bassarab S. 1984. Planta (in press)
4. Werner D, Mellor RB, Hahn MG, Grisebach H. 1984. Planta (submitted)

THE ISOENZYMES OF GLUTAMINE SYNTHETASE IN *PHASEOLUS VULGARIS*: COMPARISON OF NODULE AND ROOT FORMS ON THE GENE LEVEL.

C. GEBHARDT, J.V. CULLIMORE*, R. SAARELAINEN and B.J. MIFLIN.
Rothamsted Experimental Station Biochemistry Department, Harpenden Herts. AL5 2JQ. U.K.
*University of Warwick, Department, of Biological Sciences, Coventry, CV4 7AL. U.K.

1. INTRODUCTION

In higher plants glutamine synthetase (GS) is an octomeric enzyme which occurs in different isoforms. In the legume *Phaseolus vulgaris* (French bean) two forms can be distinguished in leaves (chloroplastic and cytosolic), one in roots and two in root nodules, one of which is induced during nodulation and seems to be nodule specific (1,2).

A cDNA-clone has been isolated from nodule poly A^+ RNA coding for a GS-subunit present in nodules but not in non-nodulated roots or leaves. According to Southern analysis there are at least four genes encoding GS-subunits (3).

2. RESULTS

A cDNA library was constructed from root poly A^+ mRNA using a novel method of synthesis of double-stranded cDNA (4). According to this method a higher proportion of full length cDNA-clones can be obtained. This root cDNA library (ca. 9500 clones) was probed with the nodule GS-clone at moderate hybridization stringency (equivalent to about 75% homology) and ca. ten clones have been identified. Four out of these ten clones are further characterized at present. They can be discriminated from the nodule-GS cDNA-clone by stringent hybridization conditions and restriction mapping whereas a discrimination between the root-GS cDNA clones is not possible by this means. By probing genomic DNA-blots (Southern) with the root-GS and nodule-GS cDNA under stringent hybridization conditions it is possible to identify restriction fragments specifically related to either root or nodule-GS.

REFERENCES

1. M. Lara, J.V. Cullimore, P.J. Lea, B.J. Miflin, A.W.B. Johnston and J.W. Lamb. 1983. Appearance of a novel form of plant glutamine synthetase during nodule development in Phaseolus vulgaris L. Planta 157, 254-258.
2. J.V. Cullimore, M. Lara, P.J. Lea and B.J. Miflin. 1983. Purification and properties of two forms of glutamine synthetase from the plant fraction of Phaseolus root nodules. Planta 157, 245-253.
3. J.V. Cullimore, C. Gebhardt, R. Saarelainen, B.J. Miflin, K.B. Idler and R.F. Barker. 1984. Glutamine synthetase of Phaseolus vulgaris L: tissue-specific expression of a multigene family. Submitted.
4. G. Heidecker and J. Messing, 1983. Sequence analysis of zein cDNAs obtained by an efficient mRNA cloning method. Nucleic Acid Res. 77, 4891-4906.

TRANSCRIPTION OF THE SOYBEAN LEGHEMOGLOBIN AND NODULIN GENES DURING NODULE DEVELOPMENT

A. MARCKER, M. LUND, K. BOJSEN, E.Ø. JENSEN and K.A. MARCKER

The soybean Lbs are encoded in the plant genome as a small family of genes consisting of four functional, one pseudo and at least three truncated genes. Six Lb genes are arranged in two independent clusters in the soybean genome. One cluster contains four Lb genes in the order 5'-Lb_a-Lb_{c1}-ψ_1Lb-Lb_{c3}-3' with the same transcriptional polarity, while the other cluster contains two genes in the order 5'-ψ_2Lb-Lb_{c2}-3' with the same polarity. The ψ_2Lb gene is a truncated gene which lacks the fourth exon and the 3' non coding end.

The activation of the soybean Lb genes during nodule development occurs exponentially. Thus about 7-8 days after infection the Lb genes are activated sequentially during a period of a few days in the opposite order to which they are arranged in the soybean genome. However, about twelve days post infection there is a dramatic increase in the transcription rates of all the Lb genes except for the Lb_{c2} gene. Concomitantly with the amplification of Lb gene transcription some of the other nodule specific plant genes (nodulin genes) are activated. The increase in transcription of the Lb genes and the nodulin genes precedes the start of transcription of the bacterial nitrogenase gene. The biosynthesis of nitrogenase implies that a change is occurring in the metabolism of the bacteria at this stage. Thus this particular change in the metabolism of the bacteria residing within the nodule is not a prerequisite for the observed changes in the transcriptional activities of the Lb and nodulin genes. Finally all the Lb genes are active for a long period during the lifetime of the nodule. Consequently the soybean Lb genes are not regulated by a developmental gene switching mechanism as is the case for regulation of vertebrate globin genes.

Ti PLASMID VECTORS FOR PLANT GENETIC ENGINEERING

M. VAN MONTAGU and J. SCHELL

Recent progress in the understanding of crown gall induction by Agrobacterium tumefaciens allowed the construction of Ti plasmids in which oncogenic functions have been replaced by the E. coli cloning vector pBR322 [1]. Foreign genes cloned in pBR322 can easily be inserted into the T-DNA by homology recombination. This type of Ti plasmid derivatives allow systematically the transfer of the genes of interest to the genome of plants. The transformed plants are capable of normal growth and differentiation [2].

Study of the DNA sequences essential for the expression of the T-DNA-encoded genes in plants resulted in the construction of (constitutive) expression vectors [3]. These vectors, first used to express antibiotic resistance genes, have also been used to express genes which can give an economically advantageous trait to plant cells. For example, the synthesis in tobacco cells of the sweetening protein thaumatine resulted in a flavour enhancement [4]. It is likely to obtain a regulated gene expression by using the appropriate regulatory DNA sequences cloned from plant cells. Positive results were obtained in tobacco by using the 5'-flanking region of a small subunit ribulose-1,5-bisphosphate carboxylase gene from Pisum sativum [5]. By fusing a reporter gene to the DNA sequences coding for the transit peptide of the small subunit of ribulose-1,5-bisphosphate carboxylase gene it was possible to translocate neomycin phosphotransferase (encoded by the reporter gene) into the chloroplasts [6,7].

It is generally thought that the Ti plasmid cannot be used as a gene vector for monocotyledonous plants. We could demonstrate that stable T-DNA transfer as well as expression of T-DNA genes is possible in some monocots. Asparagus officinalis was used as model system from which Agrobacterium-transformed cell cultures were established [8].

References

1. Zambryski P, Joos H, Genetello C, Leemans J, Van Montagu M, Schell J. 1983. Ti plasmid vector for the introduction of DNA into plant cells without alteration of their normal regeneration capacity. EMBO J. 2, 2143-2150.
2. De Block M, Herrera-Estrella L, Van Montagu M, Schell J, Zambryski P. 1984. Expression of foreign genes in regenerated plants and their progeny. EMBO J. 3, 1681-1689.
3. Herrera-Estrella L, Depicker A, Van Montagu M, Schell J. 1983. Expression of chimaeric genes transferred into plant cells using a Ti-plasmid-derived vector. Nature (London) 303, 209-213.
4. Revets H. et al. in preparation.
5. Herrera-Estrella L, Van den Broeck G, Maenhaut R, Van Montagu M, Schell J, Timko M, Cashmore A. 1984. Light-inducible and chloroplast-associated expression of a chimaeric gene introduced into Nicotiana tabacum using a Ti plasmid vector. Nature (London) 310, 115-120.
6. Van den Broeck G, Timko MP, Kausch AP, Cashmore AR, Van Montagu M, Herrera-Estrella L. 1984. Targeting foreign polypeptides to a specific organelle : translocation of the bacterial neomycin phosphotransferase (II) into the chloroplasts of transformed plant cells. Submitted.
7. Schreier P, Setfor E, Schell J, Bohnert H. 1984. The use of nuclear encoded sequences to direct the light-regulated synthesis and transport of a foreign protein into plant chloroplasts. Submitted.
8. Hernalsteens J-P, Thia-Toong L, Schell J, Van Montagu M. 1984. An Agrobacterium-transformed cell culture from the monocot Asparagus officinalis. Submitted.

ASPECTS OF THE APPLICATION OF IN VITRO (DNA) TRANSFORMATION SYSTEMS IN PLANTS.

F.A. KRENS, R.PEERBOLTE, J.H.C. HOGE, G.J. WULLEMS and R.A. SCHILPEROORT

Department of Plant Molecular Biology, State University of Leiden, the Netherlands.

1. INTRODUCTION

Based on the interaction of the soil bacterium Agrobacterium tumefaciens with its host plants, a few methods are now available for the introduction of genetic information in plant cells. These methods include in vivo infection, in vitro cocultivation and in vitro DNA transformation. The two in vitro methods have been developed in our laboratory (1, 2) and experiments have been done to investigate the potentials of both methods. Here a comparison is made between the two in vitro methods for their applicability in the genetic manipulation of plants.

2. RESULTS AND DISCUSSION

2.1. T-DNA organization in transformants

Upon cocultivation, the T-DNA generally has a fixed size bordered by the two border sequences of the Ti plasmid. This T-DNA is usually present at low copy number and is transmitted to progeny plants in a Mendelian fashion.

After in vitro DNA transformation, the T-DNA is present in various forms, i.e. as Ti plasmid derived fragments of different lengths integrated at different sites in the plant genome. The copy number of the transferred DNA fragments varies from one to several copies. The scrambled nature of the transferred DNA is reflected in the transcript patterns observed for pTi T-region fragments. An important conclusion from the DNA transformation experiments is that the Ti plasmid border sequences are not necessary for the integration of foreign DNA in the plant genome. This is also indicated by experiments in which cloned pTi T-region fragments were used as donor DNA. In this case too, integration has been found.

2.2. Co-transformation

If one wants to introduce two functions in plants (one being a selectable marker, the other a desired gene) via cocultivation experiments, these two properties have to be introduced between the pTi border sequences. The DNA transformation experiments however have shown that also pTi fragments located outside the border sequences (e.g. from the Vir-region) can become integrated. Even calf thymus DNA, used as carrier DNA in DNA transformation experiments, has been shown to be present in some transformants. This implies that any DNA, like for instance *E. coli* derived vectors, carrying genes of interest can be integrated and that it should be possible to introduce genes which are located on different plasmids by means of co-transformation.

2.3. T-DNA in progeny plants

After cross pollination with wild type pollen, one particular tobacco transformant transmitted the crown gall phenotype to 50 % of its offspring. Interestingly, also seedlings with a wild type morphology appear to contain pTi-DNA and calf thymus DNA sequences. This indicates integration of several pTi-DNA sequences as separate units in different plant chromosomes. Apparently one such pTi-DNA unit contains an active crown gall 'oncogene' which seggregates in a Mendelian fashion to 50 % of the offspring. The finding that some seedlings contain only pTi-DNA sequences and others only calf thymus DNA sequences might indicate that the calf thymus DNA is integrated independently from the pTi-DNA.

2.4. Transformation frequency

The transformation frequency of DNA transformation is about 10^{-5} which is considerably lower than for cocultivation. One has to keep in mind however that these frequencies are based on the number of transformants that show a crown gall phenotype. Transformed tissues, indeed containing foreign DNA but lacking (an) active 'oncogene(s)', do not survive selection and will not contribute to the transformation frequency.

REFERENCES

1. Marton, L., Wullems, G.J., Molendijk, L. and Schilperoort, R.A. Nature 277: 129-131 (1979).
2. Krens, F.A., Molendijk, L., Wullems, G.J. and Schilperoort, R.A. Nature 296: 72-74 (1982).

HAIRY ROOT: T-DNA, T-DNA TRANSCRIPTS AND PROGENY.

R. PEERBOLTE, J.H.C. HOGE, G.J. WULLEMS and R.A. SCHILPEROORT
Department of Plant Molecular Biology, State University of Leiden, the Netherlands.

1. INTRODUCTION

Strain LBA 9402 of the soil bacterium Agrobacterium rhizogenes can introduce (a) piece(s) of its Ri plasmid into cells of dicotyledonous plants. Transformed cells differentiate into hairy roots which can be brought into tissue culture and transformed plants may be regenerated from hairy root tissue. Some phenotypic traits of these hairy root plants are : (1) an excessive root system; (2) wrinkled leaves; (3) reduced male fertility; (4) their cells contain agropine and mannopine.

Studies on hairy root tissues of N. plumbaginifolia (RNP1), N. tabacum (HR-SR1-9402) and Daucus carota (AKR828) are presented here.

2. RESULTS AND DISCUSSION

2.1. Analysis of T-DNA and T-DNA transcripts

2.1.1. RNP1. From RNP1, an agropine and mannopine containing N. plumbaginifolia hairy root tissue induced by LBA 9402, DNA and poly(A)-RNA were isolated and analysed by Southern and Northern blot hybridization, respectively. The tissue appears to contain both T_L- and T_R-DNA. The T_L-DNA codes for at least ten transcripts and the T_R-DNA for at least two. The transcripts range in size from about 600 to 2600 nucleotides. Location and polarity of the transcripts have been determined. None of the hairy root T-DNA transcripts appeared to hybridize to octopine crown gall T-DNA probes under stringent conditions.

2.1.2. AKR828. From LBA 9402 induced Daucus carota hairy root tissue which does not contain agropine nor mannopine, DNA and poly(A)-RNA were isolated and analysed as well. This tissue does contain T_L-DNA and T_L-DNA transcripts, but lacks T_R-DNA and hence T_R-DNA transcripts. This indicates that genes coding for agropine and mannopine synthesis are located on T_R-DNA and that T_L-DNA harbours hairy root 'oncogenes'.

2.2. Progeny of hairy root plants

2.2.1. RNP1. RNP1-plants are male sterile. After cross pollination with wild type *N. plumbaginifolia* pollen, 30 - 40 % of the seedlings show all hairy root characteristics, including the presence of agropine, mannopine and (according to dot blot hybridization) T_L- and T_R-DNA. The rest of the seedlings show a wild type phenotype. Apparently there is no significant seggregation of T_L-DNA and T_R-DNA in this case.

2.2.2. HR-SR1-9402[x]. These hairy root *N. tabacum* plants show reduced male fertility. After emasculation and cross pollination with wild type SR1 pollen, half of the resulting seedlings show a hairy root phenotype (excessive root system, wrinkled leaves). The other half has a normal appearance. Dot blot hybridization revealed that all hairy root looking seedlings contain T_L-DNA; normal looking seedlings do not. Surprisingly not all hairy root looking seedlings contain agropine and mannopine. Those that contain agropine and mannopine also contain T_R-DNA sequences. One half of the normal looking seedlings appears to contain agropine, mannopine and T_R-DNA. So here seggregation of T_L- and T_R-DNA is found. Again these results indicate that T_L-DNA contains hairy root 'oncogenes' and that genes necessary for agropine/mannopine synthesis are encoded by T_R-DNA.

When a nopaline crown gall *N. tabacum* graft (lacking an active crown gall auxin locus) is cross pollinated with hairy root *N. tabacum* pollen, half of the progeny has crown gall features, one quarter has a hairy root phenotype and one quarter looks normal. This indicates that the crown gall cytokinin locus dominates the excessive root system induced by hairy root T_L-DNA.

x: kindly provided by L. Spano, Italy.

HAIRY ROOT: POTENTIAL FOR GENETIC ENGINEERING AND PLANT CELL PHYSIOLOGY

P.Costantino,M.Cardarelli,M.L.Mauro,M.Pomponi and L.Spanò
Dipartimento di Genetica e Biologia Molecolare, Università di Roma,Roma,Italy
Laboratori ANFA,ASSORENI,Monterotondo,Roma,Italy
Centro Acidi Nucleici C.N.R.,Roma,Italy

Agrobacterium rhizogenes is responsible for the plant disease hairy root (1),characterized by abudant proliferation of roots at the site of bacterial infection.

The virulence properties of A.rhizogenes are associated with the presence of large plasmids (98 to 190 Mda) denominated Ri (root-inducing) plasmids (2,3). Recently it has been demonstrated that a portion of these plasmids (T-DNA) is transferred and integrated in the infected plant cell genome (4,5,6,7). The presence of T-DNA somehow affects plant cell growth and induces differentiation into roots.

Hairy root cells synthesize unusual aminoacid derivatives called opines of which two major classes have been so far characterized: mannopine and its lactonized derivative agropine (8,9). The two classes of opines correspond to two families of Ri plasmids whose members are closely related with each other (10).

Ri plasmids have been compared with the better characterized Ti plasmids of A.tumefaciens and have been shown to share with these latter grossly the same general organization (11,12).

Substantially different seem to be however the T-DNAs of Ri and Ti plasmids. Thus, despite hairy root and crown gall tumours share many common features at the induction level and basically analogous molecular transformation events, the actual genes responsible for plant cell proliferation seem to be different in the two bacterial plasmids (11,12). On the contrary very similar with each other are the T-DNAs of mannopine and agropine Ri plasmids (7). While the former T-DNA is however a single continuous segment of DNA some 15 to 30 Kilobases long (13), the agropine T-DNA Consists of two stretches of DNA (TL and TR)separated by a long plasmid sequence which is not integrated in the plant genome (14). The TL-DNA harbours the functions necessary for induction and maintenance of hairy roots and is homologous to the single T-DNA of mannopine Ri plasmids (7). The TR-DNA has been very recently demonstrated to harbour all of the genes that encode agropine biosynthesis in transformed plant cells (14). The presence of two separate segments of T-DNAs in agropine hairy roots might be of practical importance in developing vectors for plant genetic engineering. The TR-DNA seems in fact to be a transforming DNA naturally devoid of any oncogenic function, equivalent to a "disarmed" Ti plasmid

T-DNA. In addition the presence and expression of Ri plasmid T-DNA does not seem to affect the regenerative capabilities of transformed cells. Hairy root cultures of several plant species have been in fact successfully induced to regenerate morphologically normal, fertile plants whose offspring still contain T-DNA (4,15,16,17).

The A.rhizogenes system offers thus formidable opportunities both for plant genetic engineering and in approaching the molecular basis of plant differentiation. We are currently involved in the construction of Ri plasmid derived vectors, devoid of the oncogenic TL-DNA, and in the characterization of the products of expressionr of the T-DNA in hairy root tissues.

REFERENCES:

(1) C.Elliot,"Manual of Bacterial Plant Pathogens" 2nd rev.ed.,Chronica Botanica,Waltham Mass.(1951)
(2) L.Moore et al. Plasmid 2:617 (1979)
(3) F.F.White and E.W.Nester J.Bacteriol. 141:1134 (1980)
(4) M.D.Chilton et al. Nature 295:432 (1982)
(5) F.F.White et al. Proc.Natl.Acad.Sci.USA 79:3193 (1982)
(6) L.Willmitzer et al. Mol.Gen.Genet. 186:16 (1982)
(7) L.Spanò et al. Plant Mol.Biol. 1:291 (1982)
(8) A.Petit et al. Mol.Gen.Genet. 190:204 (1983)
(9)M.E.Tate et al. Carbohydrate Res. 104:105 (1982)
(10) P.Costantino et al. Plasmid 5:170 (1981)
(11) F.F.White and E.W.Nester J.Bacteriol. 144:710 (1980)
(12) G.Risuleo et al. Plasmid 7:45 (1982)
(13) M.Byrne et al. J.Mol.Appl.Genet. 2:201 (1983)
(14) A.De Paolis et al. submitted (1984)
(15) L.Spanò and P.Costantino Z.Pflanzenphysiol. 106:87 (1982)
(16) P.Costantino et al. J.Mol.Appl.Genet. in press (1984)
(17) C.David et al. Biotechnilogy 2:73 (1984)

GENE FUSIONS WITH T-DNA FRAGMENTS OF AGROBACTERIUM RHIZOGENES ARE EXPRESSED IN E. COLI.

P. RATET, F. BOULANGER, A. BEYOU, F. MIGNOTTE, A. COMBARD, A. BERKALOFF and F. RICHAUD.

Agrobacterium rhizogenes transfers a part of its plasmid (T-DNA) to the genome of plant upon infection. One particular part is named c-TDNA because homology with T-DNA of nontransformed plant was observed.
Using a miniMu(Km,*lac*') transposable element able to form translational fusions of the *lacZ* coding region with the gene in which the insertion has occured ; 45 independent insertions in the T-DNA were obtained showing different levels of expression of *B*-galactosidase ; 20 of them are localized in the c-TDNA part.
This result suggest that at least some of the genetic information of the *A. rhizogenes* T-DNA is able of expression in *E. coli*.
The nucleotidic sequences responsible for the expression of several of these inserts localized in the c-TDNA has been determined ; several open reading frames were identified.
20 inserts were transformed in *Agrobacterium*. The resulting insertion mutagenized strains were used to infect carrot slices and different phenotypes were obtained. The transformed roots are grown to study *B*-galactosidase expression.

A new miniMu(Cm,Km') was engineered *in vitro*. This transposable element is able to form translational fusions of the Km' coding region with the gene in which the insertion has occured. Several insertions in the T-DNA were obtained showing a Km resistance phenotype in *E. coli*. These inserts are transformed in *Agrobacterium* and the resulting strains will be used to infect plants and study the resistance to Km of the transformed plants.

REFERENCES

Castilho B, Olfson P and Casadaban M. (1984) Plasmid insertion mutagenesis and *lac* gene fusion with mini-Mu bacteriophage transpóson. J. Bacteriol. *158*, 488-495.
Huffman G, White F, Gordon M and Nester E. (1984) Hairy-root inducing plasmid : physical map and homology to tumor-inducing plasmids. J. Bacteri(*157*, 269-276.

THE FUNCTION OF T-DNA IN NATURE

Jacques Tempé and Annik Petit

GRIMP-INRA, LA136-CNRS
Institut de Microbiologie,
Bâtiment 409, Université de Paris-Sud, F-91405 ORSAY.

The finding that crown gall tumors, elicited by *Agrobacterium tumefaciens* strains harbouring a pathogenic Ti plasmid, contain new and specific compounds, now called opines, has been one of the milestones in the development of crown gall research. The demonstration that the crown gall cell phenotype, including opine synthesis, is coded for by Ti plasmid DNA sequences integrated into the genome of plant cells has opened the way to genetic engineering of plant cells mediated by Ti plasmid-derived vectors. Opine synthesis is now currently used as a distinctive marker for engineered plant cells.

Besides the development of plant genetic engineering, the study of the pathogenic plasmids of *Agrobacterium* may be responsible for progresses in the understanding of interactions between plants and bacteria (1). Thus the concept that opine synthesis in crown gall tumors serves the purpose to create an ecological niche for pathogenic *Agrobacterium* strains which has been proposed to account for the already ancient observation of a correlation between opine synthesis in crown gall tumors and Ti plasmid-coded opine degradation by *Agrobacterium* strains (2) has led to interesting generalizations. In short, this theory states that opines synthesized by transformed plant cells are used as selective growth substrates by the pathogen and that the availability of opines is a condition for its propagation (3). If this is true, opine-related functions must be essential for the survival of Ti plasmids and, consequently, any one of them should carry such functions.

We have looked for opine-related functions in every *A. tumefaciens* strain that had not yet been characterized with respect to such functions. These investigations have resulted in the discovery of several new opines and the confirmation that every strain and plasmid studied carries opine-related functions (4,5). Parallel studies on *A. rhizogenes* and their Ri plasmids, which elicit the formation of rooty proliferations called hairy-root on dicotyledonous plants, have shown that the opine concept also applies to this system (6). At the present time, we know of no exception to the rule that *Agrobacterium* pathogenic plasmids carry opine related functions and, besides the two original Ti plasmid opine-types (octopine-type and nopaline-type) that constituted the basis on which the opine concept was formulated, 6 new Ti or Ri plasmid opine-types have now been identified. We, therefore, feel confident that the opine concept must describe rather accurately the role of opines in the interaction between *Agrobacterium* strains, their pathogenic plasmids and the host plants and we conclude that the function of T-DNA in nature is to provide, via opine synthesis, a nutritional advantage to the pathogen that is necessary for its survival.

The discovery of opine-like substances in some legume nodules (7) raises the question whether the opine concept applies to other plasmid-mediated plant-bacteria interactions.

REFERENCES

1. Tempé, J., Petit, A. (1983). La piste des Opines. In Molecular genetics of the bacteria-plant interaction, A. Pühler Ed., Springer-Verlag, Berlin Heidelberg, 14-32.

2. Petit, A., Delhaye, S., Tempé, J., Morel, G. (1970). Recherches sur les guanidines des tissus de crown gall. Mise en évidence d'une relation biochimique spécifique entre les souches d'Agrobacterium tumefaciens et les tumeurs qu'elles induisent. Physiol. vég., **8**, 205-213.

3. Petit, A., Dessaux, Y., Tempé, J. (1978). The biological significance of opines. I. A study of opine catabolism by Agrobacterium tumefaciens. Proc. 4th Inter. Conf. Plant Pathog. Bact., M. Ridé, Ed., INRA - Angers, 143-152.

4. Guyon, P., Chilton, M.D., Petit, A., Tempé, J. (1980). Agropine in "null type" crown gall tumours : evidence for the generality of the opine concept. Proc. Natl. Acad. Sci. U.S.A., **77**, 2693-2697.

5. Chilton, W.S., Tempé, J., Matzke, M., Chilton, M.D. (1984). Succinamopine a new crown gall opine. J. Bacteriol. **157**, 357-362.

6. Petit, A., David, C., Dahl, G.A., Ellis, J.G., Guyon, P., Casse-Delbart, F., Tempé, J. (1983). Further extension of the opine concept: Plasmids in Agrobacterium rhizogenes cooperate for opine degradation. Mol. Gen. Genet. **190**, 204-214.

7. Tempé, J., Petit, A., Bannerot, H. (1982). Présence de composés semblables à des opines dans des nodosités de luzerne. C.R. Acad. Sci. Paris, Série III, **295**, 432-434.

THE USE OF T-DNA GENES TO MODIFY POTATO DEVELOPMENT.

G. OOMS, M.M. BURRELL, A. KARP, J.R. LENTON*, J. ROBERTS and D. TWELL.
Rothamsted Experimental Station, Harpenden, Herts., U.K.
* Long Ashton Research Station, Bristol, U.K.

Tumour development on dicotyledonous plants infected with Agrobacterium tumefaciens is the result of the stable integration into a plant chromosome and the expression of a segment of a large tumour inducing (Ti-) plasmid. This transferred DNA or T-DNA carries genes whose expression products influence the concentration of endogenous plant hormones. Collectively the genes enhance auxin and cytokinin concentrations causing transformed cells to grow autonomously even when cultured in vitro, away from the plant and the inducing bacteria.

On the one hand, removal of the hormone related T-DNA genes has led to the construction of advanced vectors for introducing genes into plant cells On the other hand, the hormone related T-DNA genes themselves are of interest as tools to modify and study, in a specific way, growth and development of plants transformed either with wild-type T-DNA genes and/or engineered derivatives of them. Tobacco plants e.g. regenerated from tumours and transformed with Ti T-DNA, usually did not form roots. Similarly, tobacco plants regenerated from hairy-roots and transformed with Ri T-DNA, from the related bacterium A. rhizogenes, have crinkled leaves and an abundant root system. Therefore, if whole plant genetic manipulation experiments were successful for plants other than the 'model' plants tobacco or petunia, it would be of particular interest to use T-DNA genes to genetically modify plants such as potato, with a well defined developmental process like tuberisation.

Using wild-type and intact A. tumefaciens and A. rhizogenes strains we have transformed explants and protoplasts of a number of tetraploid potato cultivars. In some cases we have regenerated from these whole transformed plants.

Shoot inducing A. tumefaciens strains caused the development of tumours on infected shoot cultures from which spontaneously shoots developed. Some of these shoots were transformed. They contained Ti

T-DNA and produced opines. Like similarly transformed tobacco plants they normally failed to form roots but via grafting they could be regenerated into mature plants. Interestingly, the grafted plants had a strong tendency to tuberise. This was correlated and most likely caused by the presence of an approximately 3-10 fold enhanced concentration of cytokinins in the grafted plants compared with non tuberising grafted control plants. The transformed potato shoots grown in vitro did not form tubers. Such shoots contained approximately 100-200 fold more biologically active cytokinins than untransformed shoots. Associated with the reduction in cytokinin concentration of transformed grafted plants vs. transformed shoots was an approximately 5 fold reduction in the expression of the T-DNA cytokinin gene.

Potato plants regenerated from hairy roots contained Ri T-DNA. Such plants, like similarly transformed tobacco plants, also formed crinkled leaves and an abundant root system. In contrast with the morphologically apparently normal tubers transformed with Ti T-DNA, tubers from Ri T-DNA transformed plants were longer and had more prominent and abundant 'eyes'.

These results have established that transformed derivatives of tetraploid potato cultivars can be isolated. They will form tubers and therefore other genes can be cointroduced into potato via the T-DNA of either Ri or Ti plasmids and their expression can be studied. However, more importantly, the results obtained have established that T-DNA genes can be used to specifically alter growth and development of potato, thereby opening the prospect of studying and modifying many aspects of plant development and in particular tuberisation in a novel way.

ONTOGENETIC PLASTICITY AND COMPETENCE FOR CROWN GALL TRANSFORMATION IN NICOTIANA spp.

Patrizia Bogani, Marcello Buiatti, Lambert A.M. Hensgens, Priscilla Bettini, Maria G. Pellegrini and Aniello Scala.

Much is known about the contribution of *Agrobacterium tumefaciens* to crown gall induction in plants. The conserved T-DNA which becomes integrated in the genome of the plant cell contains genes encoding enzymes which catalyse the synthesis of endogenous growth regulators in the transformed plant cells. However little is known about the influence of the plant cell on the process of tumor transformation by *A. tumefaciens*. The *Nicotiana* genus may be a suitable genetic system to study the contribution of the plant genome to tumor induction. Hybrids, formed between species belonging to different groups ("plus" and "minus"), form spontaneously tumors (1). The formation of these tumors has been correlated with differences in the plants capacity to synthesize growth regulators (2).

Differentiation and tumor formation were analysed in 9 *Nicotiana* species, the tumorous amphidiploid *N. glauca* x *N. langsdorffii* hybrid and a nontumorous mutant of it. In a first serie of experiments, calluses of all tested genotypes, grown for two months on a Linsmayer and Skoog medium supplemented with 2,4-D, were transferred to the same medium without hormone. Shoot- and root-formation and autonomous growth were scored at different times.

As far as "plus" species were concerned those belonging to the Alatae showed good shoot differentiation ability. *N. langsdorffi* calluses produced plantlets while *N. plumbagini-*

folia, N. longiflora, N. bigelovii var. quadrivalvis, N. bigelovii var. bigelovii and the tumorous amphidiploid were all able to grow without hormones. "Minus" species (N. tabacum, N. knightiana, N. paniculata and N. glauca) either formed roots or did not differentiate. None of these four species could be habituated. The nontumorous N. glauca x N. langsdorffii did not differentiate nor showed autonomous growth.

To investigate a possible correlation between the differentiation behaviour and competence for crown gall induction, cuttings of aseptically grown plants of the described genotypes were infected with A. tumefaciens B6 and scored for tumor formation. All genotypes formed tumors. The nontumorous mutant however developped outgrowths, which on a hormone free medium eventually died. N. langsdorffii was particularly late in tumor formation; tumor induction however could be accelerated in this genotype as in the nontumorous mutant by kinetin and IAA pretreatment of the cuttings.

In general "habituation prone" species were fast in tumor induction and formed larger tumors while "minus" species and N. langsdorffii seemed to be less easily transformed. An exception to this pattern is N. glauca, which formed large tumors after a short lag time. Experiments are now in progress to see whether T-DNA integration and expression differ between the studied genotypes.

REFERENCES

1. Näf, U. 1958. Growth 22; 167-180.
2. Buiatti, M. and Bennici A. 1970. Proc. Accad. Naz. Lincei 48; 261-269.

RNA VIRUSES AS VECTORS IN GENETIC ENGINEERING OF PLANTS

JOHN VAN EMMELO, PAUL AMELOOT AND WALTER FIERS

1. INTRODUCTION

In order to evaluate the possible use of RNA viruses as vectors for the transfer of heterologous information, we decided to develop a model vector system with Satellite Tobacco Necrosis Virus (STNV) as a vehicle using the potent and controllable procaryotic transcription-translation system derived from bacteriophage (Remaut et al., 1981). The main objective is the production of infectious plant viral components in E. coli in such a way that subsequent alterations at the DNA (plasmid) level would be reproduced in the viral genome.

2. RESULTS

A full-size cDNA copy of STNV-RNA placed behind the very efficient and controllable bacteriophage - pL promoter allows the synthesis of a substantial amount of STNV coat protein (van Emmelo et al., 1984). Unfortunately, endogenous degradation systems of E. coli rapidly degrade the newly synthesized proteins. Although this could be partially circumvented, we decided to focus on the transcription process because preliminary experiments clearly indicated that primary leaves of Cowpea could, with a considerable lower efficiency as compared to virus particles, be mechanically infected with non-encapsidated RNA molecules.

RNA-transcripts were isolated from temperature-induced E. coli cells and, after denaturation, separated on agarose gels. After transfer to nitrocellulose membranes ("Northern blotting"), STNV information containing RNA was detected using radioactively labelled STNV-DNA probes. Surprisingly, it was demonstrated that the major STNV-transcription product was only slightly larger than infectious STNV-RNA. As all the transcripts contain a 150 nucleotide plasmid-derived leader sequence at their 5'-end, we could conclude that the 3'-end of the transcripts is situated very near to the 3'-end of the STNV-information. Data, showing that the surrounding plasmid region did not influence this situation, seem to confirm this conclusion. Furthermore, we were able to demonstrate that this phenomenon was

N-protein-resistant.

We are presently unable to differentiate between the presence of a real transcriptional stop or post-transcriptional processing to explain our findings. Hybridization with well defined, non-STNV-dependent probes, as well as the use of RNAase III-deficient mutants will hopefully resolve this problem.

3. CONCLUSION

Synthesis of both STNV capsid protein and viral RNA in E. coli, programmed with expression vectors containing a full-size DNA copy of STNV-RNA, seem to give rise to unexpected problems. Not only are the in vivo synthesized coat protein molecules rapidly degraded by intracellular regulatory protease systems, but the STNV-information itself contains a signal for termination of transcription or for post-transcriptional processing so that full-size (containing the complete 3'-end information) transcipts are difficult to demonstrate. In vitro transcription systems are now being applied to resolve the difficulties.

REFERENCES

1. REMAUT E, STANSSENS P & FIERS W. 1981. Plasmid vectors for high-efficiency expression controlled by the P_L promoter of coliphage lambda. Gene 15, 81-93.
2. VAN EMMELO J, AMELLOT P, PLAETINCK G & FIERS W. 1984. Controlled synthesis of the coat protein of Satellite Tobacco Necrosis Virus in E. coli. Virology 136, 32-40.

HYBRID VECTORS CONTAINING A PROMOTOR FROM CAULIFLOWER MOSAIC VIRUS AND PLANT AUTONOMOUSLY REPLICATING SEQUENCES (ARS) SELECTED IN YEAST : ATTEMPTS TO TRANSFORM PROTOPLASTS.

E. BALAZS, S. BOUZOUBAA, H. GUILLEY, G. JONARD, K. RICHARDS & L. HIRTH
IBMC, 15 rue Descartes, 67000 Strasbourg, France

An expression vector capable of autonomous replication in plant cells would be a useful complement to the integrative vectors based on Ti plasmid which are currently used to introduce foreign genes into plants. It has recently been shown by Brisson et al. (1) that cauliflower mosaic virus can be employed as such a vector by replacing the nonessential open reading frame II with the dihydrofolate reductase gene. The principal limitations of this vector stem from (i) the rather small host range of the virus, (ii) the limited space available for insertion of foreign DNA, and (iii) the need to precisely eliminate almost all noncoding sequences surrounding the insert. An approach circumventing these drawbacks would be to use cauliflower mosaic virus only as a source of initiation and termination signals for transcription of the foreign gene combined with a replication origin derived from plant DNA. Such a vector might be able to replicate autonomously under the control of the host plant's replication machinery.

Plasmids capable of correcting auxotrophy in brewer's yeast Saccharomyces cerevisiae have been used to isolate autonomously replicating sequences (ARS) from a number of organisms (2,3,4). We have used this system to isolate ARS elements from the DNA of a cruciferae, Arabidopsis thaliana, and from tobacco, Nicotiana sylvestris. Several ARS elements from A. thaliana have been sequenced and shown to be AT-rich and to possess long tracts of alternating purines and pyrimidines. In order to learn if these sequences, which permit autonomous replication of a plasmid in yeast, can fulfill a similar function in plant cells we have inserted ARS elements into a plasmid containing the neomycin phosphotransferase gene from Tn5 under control of the gene VI promotor from cauliflower mosaic virus and used these constructs to transform protoplasts to resistance to the neomycin analogue G418. With a polyethylene glycol transformation procedure we were able to generate microcalli from turnip protoplasts resistant to 40 µg/ml G418, a concentration lethal to mock-transformed protoplasts. The resistant microcalli typically survived 3-4 months (4-5 transfers), grew to several mm diameter and then died or stopped growing. Similar results were obtained whether or not an ARS element was present in the transforming plasmid, suggesting that the putative transformants probably occurred via insertion of the plasmid DNA into the plant genome rather than by its autonomous replication. An experiment to detect plasmid DNA sequences in the DNA from G418-resistant microcalli by Southern hybridization, although hampered by the small amounts of tissue available, gave a very faint band of intensity well below that expected for one copy per genome equivalent of plant DNA, suggesting that the microcalli may be mosaics containing many nontransformed cells. Attempts are now underway to perform similar transformation experiments with tobacco protoplasts.

REFERENCES

1. Brisson N, Paszkowski J, Penswick JR, Gronenborn B, Potrykus I, Hohn T. 1984. Expression of a bacterial gene in plants by using a viral vector. Nature 310, 511-514.
2. Stinchcomb DT, Stuhl K, Davis RW. 1979. Isolation and characterization of a yeast chromosomal replicator. Nature 282, 39-43.
3. Kearsey S. 1983. Analysis of sequences conferring autonomous replication in baker's yeast. EMBO J. 2, 1571-1575.
4. Montiel JF, Norbury CJ, Tuite MF, Dobson MJ, Mills JS, Kingsman AJ, Kingsman SM. 1984. Characterization of human chromosomal DNA sequences which replicate autonomously in Saccharomyces cerevisiae. Nucleic Acids Res. 12, 1049-1068.

GENETIC CHARACTERIZATION OF ECTOMYCORRHIZAL FUNGI.

F.MEINHARDT and K.ESSER
Lehrstuhl für Allgemeine Botanik, Ruhr-Universität,
Postfach 102148, D-4630 Bochum, Germany.

INTRODUCTION

For any genetic work it is a prerequisite to characterize the organisms concerned. For genetic manipulation of fungi extrachromosomal DNA such as plasmids and/or mitochondrial (mt) DNA is of special importance (Esser et al. 1983).

We have therefore looked for plasmids in several groups of ectomycorrhiza forming fungi, including the ascomycetous genera *Morchella* and *Gyromitra* as well as the Basidiomycetes *Boletus* and *Pisolithus*.

RESULTS

After having detected linear plasmids in *Morchella conica* previously (Meinhardt and Esser 1984), other strains of Morchella were screened for comparable genetic traits. Linear plasmids with a molecular weight in the range from 6 to 8 kbp could be detected in *M.conica* and *M.hortensis*. Among the Basidiomycetes in one case a linear plasmid of 7.7 kbp was detected. RNAse, DNAse digestions and restriction analysis confirmed the DNA nature and the linearity of the molecules.

In addition the mt DNA of some strains was isolated and characterized preliminary by restriction analysis.

CONCLUSION

The detection of plasmids in mycorrhiza forming fungi allows the construction of vectors capable of replication in this group of fungi and may lead perhaps to the answer of the question: How do symbionts communicate?

REFERENCES

Esser K, Kück U, Stahl U, Tudzynski P. 1983. Cloning vectors of mitochondrial origin for eukaryotes: a new concept in genetic engineering. Curr.Genet. 7, 239-243.

Meinhardt F, Esser K. 1984. Linear extrachromosomal DNA in the morel Morchella conica. Curr.Genet. 8, 15-18.

ANALYSIS OF TRANSFORMATION EVENTS IN ASPERGILLUS NIDULANS

P. DURRENS, P.M. GREEN, H.N. ARST Jr. and C. SCAZZOCCHIO

1. INTRODUCTION

Many filamentous fungi are of industrial importance for they can produce high value added molecules (i.e. enzymes, antibiotics). Thus it is of interest to establish procedures enabling artificial modifications of fungal genomes.

Aspergillus nidulans was transformed by integration with a plasmid containing the L-proline catabolism gene cluster (prn). DNAs of transformants and their progeny were investigated.

2. THE PROLINE CATABOLISM GENE CLUSTER OF A. NIDULANS

The prn cluster, located in the linkage group VII, has the gene order prnA (a positive acting regulatory gene) - prnD (proline oxidase gene) - a cis acting control region - prnB (major proline permease gene) - prnC (Δ^1-pyrroline-5-carboxylate dehydrogenase gene). (1).

3. TRANSFORMATION

A derivative of pBR322 plasmid, containing the almost entire prn cluster, was used to transform prn-deleted strains (2) by the method of Tilburn et. al. (3).

4. ANALYSIS OF TRANSFORMATION EVENTS

Transformants from Δ353 and Δ309 deletion strains were investigated. Deletion Δ353 includes prnA - prnD - control region and prnB and deletion Δ309 covers prnD and control region.

Localisation of plasmid insertion sites and selection of transformant-derived strains were obtained by formal genetical analysis.

4. 1. Transformants from Δ309 strain

In transformant T1, multiple copies of the plasmid were inserted in linkage group V (i.e. a non-homologous region). The multiple inserts led to a derepressed proline utilisation phenotype. Analysis of transformant DNA by Southern blot hybridisation with cloned prn fragment or pBR322 probes suggested an internal recombination between the inserted plasmids. Although mitotically stable, inserts could be eliminated through meiosis: a strain containing a single copy of plasmid only was found in progeny arising from self fertilisation.

The transformant T2 had the same phenotype as T1 but the insertions occured in the normal chromosomal prn location. New bands on Southern blot of transformant T2 DNA indicated the generation of a new distinct deletion. Elimination of inserts via selfing led to a strain exhibiting a mutant phenotype resulting from the new deletion.

4. 2. Transformant from Δ353 strain

Here, a small number of inserts at the prn site resulted in a wild-type proline utilisation phenotype. But a display of new bands on Southern blot of transformant DNA indicated once more the presence of a new deletion. Formal genetical analysis gave the evidence that the two deletions have a common endpoint or a short overlapping region.

5. CONCLUSION

The insertion of DNA into *Aspergillus* genome can occur even in non-homologous sites as seen previously in transformation with the amdS - containing plasmid (3 and unpublished results).

The prn-plasmid inserts appear to be quite unstable through meiosis.

The transformation of deleted strains seems to be an easy way of obtaining new deletions of genome regions at specified sites.

REFERENCES

1. Jones SA. et al. 1981. Genes roles in the prn cluster of *A. nidulans*. Current Genetics 3, 49-56.
2. Green PM 1984. Thesis, University of Essex, England.
3. Tilburn J et al. 1983. Transformation by integration in *A. nidulans*. Gene 26, 205-221.

THE IRON UPTAKE SYSTEM OF PLANTGROWTH-STIMULATING PSEUDOMONAS

J. Marugg, M. van Spanje, B. Schippers and P.J. Weisbeek
Department of Molecular Cell Biology, Department of Phytopathology and Institute of Molecular Biology, University of Utrecht, Utrecht, The Netherlands

INTRODUCTION

Fluorescing Pseudomonads can be isolated from roots of certain plants that are beneficial to the plant (e.g. potato, radish, sugarbeet). Treatment of seeds of these plants with such bacteria causes a significant increase in crop yield (1). The Pseudomonas species posses very effective systems for aquisition of iron from the soil and when growing under iron-limiting conditions, they inhibit the growth of several types of pathogenic micro-organisms. We are investigating the relation between the presence of such iron-uptake systems and the growth-promoting activity of these bacteria.

RESULTS

The Pseudomonas putida strain WCS358, isolated from potato-roots, fluoresces strongly when grown under iron-limiting conditions. Concentrations of up to 800 μgr/ml 2,2-bipyridyl do not interfere with the growth of WCS358 indicating the synthesis of a siderophore with high iron-binding capacity. This strain can be transformed efficiently and conjugates well with RP4-containing E. coli cells. The latter property was used for in vivo mutagenesis of WCS358. The pBR325 derivative pSUP2021 (containing Tn5 and the mob-site of RP4) in the E. coli strain S17-1 (with immobilezed RP4 tra-genes) (2) was transferred to WCS358. Pseudomonas cells became resistant to kanamycin and streptomycin (both located on Tn5) with a frequency of 10^{-6}.
Hybridization of mutant chromosomal DNA blots with Tn5 DNA showed the presence of unique Tn5 insertions at different chromosomal locations in the various mutant strains.
The mutant bank contained ca. 2% auxotrophic mutants. Non-fluorescing mutants and mutants with altered growth characteristics under iron-limiting conditions were selected from the mutant bank. These mutants could be divided in $fluo^-sid^-$, $fluo^-sid^+$, $fluo^+sid^-$, $fluo^{\pm}sid^{\pm}$ and $fluo^{\pm}sid^+$.

The fluo$^-$ mutants were found with a frequency of ca. 2%. The fluo$^-$sid$^+$ mutants do not fluoresce but they excrete a siderophore with iron-binding properties comparable to the wild type fluorescent siderophore. The fluorescence-deficient siderophore can be taken up and used by other, different, mutants to synthesize the complete fluorescing compound (cross-feeding). These results show that the fluorescing compound is identical to the siderophore and that the cells contain only a singly major siderophore.

A cosmid colony bank of wild type WCS358 DNA in pLAFR1(3) was conjugated with the set of mutants and the complementing cosmids were identified by their ability to restore fluorescense and/or iron binding.
A minimum of 8 separate genes have been identified by this complementation system. At least 4 of these genes form a cluster on the chromosome. Work is in progress to extend the cosmid-analysis and to identify the genes involved.

REFERENCES

1. Geels, F.P., Schippers, B. 1983. Reduction of yield depressions in high frequency Potato cropping after seed tuber treatments with antagonistic fluorescent Pseudomonas spp. Phytopath. Z., 108, 207-214.
2. Simon, R., Priefer, U. and Pühler 1983. A broad host range mobilization system for in vivo genetic engineering: transposon mutagenesis in gram negative bacteria. Biotechnology, november.
3. Friedman. A.M., Long, S.R., Brown, S.E., Buikema, W.J. and F.M. Ausubel 1982. Construction of a broad host range cosmid cloning vector and its use in genetic analysis of Rhizobium mutants. Gene, 18, 289-296.

Investigation of the molecular basis of virulence in the plant pathogen Pseudomonas syringae

C. Grimm, N. J. Panopoulos

1. Introduction

The Pseudomonas group is known to include a considerable amount of pathogens which have a world-wide distribution and cause different types of diseases in host-plants. We are interested in the question by which properties pathogens are able to injure plants and what makes a plant susceptible to infection, or, on the other hand, what is the nature of plant resistence. Understanding of these processes will probably help to control microbial pathogens.
The host-pathogen interactions have been studied extensively and although molecular mechanisms are still unknown, those investigations revealed an interaction between specific genes in the pathogen and specific genes in the host-plant (Flor, 1947,1955) (Ellingboe, 1976). These gene interactions are very complex, resulting in development of disease symptoms in the host-plant when injected by homologous virulent pathogens or in an hypersensitive reaction (HR, represents an expression of plant resistence)of the plant to avirulent (unable to induce disease) or heterologous pathogens.

2. Project

We are going to investigate the molecular basis of virulence (avirulence) of members of the Pseudomonas syringae group using the available methods of molecular genetics. We have chosen the *Pseudomonas* pathovars *syringae* and *savastanoi* as objects for investigations. They cause fruit lesion on lemon and olive knots, respectively and thus influence the greek cultivation considerably. We plan to construct genomic libraries in the broard host range vector pLAFR3 and to screen these libraries for different virulence (avirulence) genes by complementation tests. We will analyse and compare these genes by restriction analysis, southern hybridization, Tn5 mutagenesis and DNA sequencing. We also intend to look for the gene product(s) which will perhaps allow us the investigation of the complementary resistence gene(s) in the host-plant.

References

Flor, H. H. 1947. Inheritance of reaction to rust in flax. J. Agric. Res. 74: 241-262.

Flor, H. H. 1955. Host-parasite interactions in flax rust-its genetics and other implications. Phytopathology 45: 680-685

Ellingboe, A. H. 1976. Genetics of host-parasite interactions. Pp. 761-778 in Encyclopedia of Plant Physiology (R. Heitefuss and P. H. Williams, eds.). Springer-Verlag.

ISOLATION OF GENES INVOLVED IN PEST AND DISEASE RESISTANCE

BOULTER, D. and EVANS, I.M.

In order to screen a pea gene bank for lectin genes we have prepared lectin cDNA clones as follows :-

Total polysomes were isolated from developing pea cotyledons at 16 days after flowering. The polysomes synthesising lectin were immunoprecipitated with specific lectin antibody and these were used as a template for the synthesis of lectin cDNA. Double-stranded cDNA was polished with DNA polymerase I, ligated to EcoR1 linkers and inserted into pUC8 vector. This cDNA bank has been screened with a number of probes, legumin (pAD4.4) vicilin (pAD3.4, pAD2.1) and a Phaseolus lectin cDNA clone and positive clones identified in each case.

The lectin cDNA clones will be used to probe pea, cowpea and bean genomic libraries which either exist or are under construction.

THE LEGUMIN GENE FAMILY OF PEA - SEQUENCES AND STRUCTURAL FEATURES

CROY, R.R.D., LYCETT, G.W., GATEHOUSE, J.A., SHIRSAT, A., LEVASSEUR, M. AND BOULTER, D.

Legumin, a major storage protein of pea (Pisum sativum L.) seeds, is encoded by a small gene family, estimated to contain 4 members by hybridisation of cloned cDNA to genomic DNA. Clones containing legumin genes have been isolated from gene libraries of pea genomic DNA, and 4 different legumin genes have been identified in these clones. Two of the genes (leg A and leg D) are arranged close together on the same genomic clone and are oriented in the same direction. Hybridisation of the gene, clones to genomic DNA shows that a few further homologous sequences (possibly legumin genes) are present. Regions flanking the legumin genes, both to 3' and 5', do not appear to contain sequences repeated in the pea genome to a high copy number.

The full sequence of the transcribed and flanking regions of an active legumin gene (leg A) is presented. The transcription start has been identified by S1 nuclease mapping and the coding sequence by comparison with protein and cDNA sequences. The gene contains three introns of approximately 80-100 bp, all of which follow the normal eukaryotic boundary rules. The coding sequence of leg A contains 3 near repeats of a 54 bp sequence, which give rise to a highly charged and hydrophilic region of amino acid sequence up to the C-terminal of the legumin α subunit. In the 5' flanking sequence 'TATA' and 'CAAT' promoter sequences are present at consensus distances from the transcription start. The 3' flanking sequence shows three separate polyadenylation signals, each of which contains two overlapping AATAAA sequences. The 3' flanking sequence also contains a putative stem-loop structure which may be involved in legumin mRNA synthesis or stability. The amino acid coding sequences of two of the other legumin genes, leg B and leg C are highly homologous to that of leg A, as estimated by restriction mapping. Data on the sequence homologies of coding, 3' and 5' flanking regions of legumin genes A, B and C will be presented.

The fourth legumin gene, leg D, which is arranged in tandem to leg A, appears to be a pseudogene, since it has a stop codon in its reading frame near the N-terminus of the potential coding sequence. In addition, it contains three deletions of approximately 65, 100 and 110 bp, one of which abolishes the 3' junction between the second intron and the following exon, and one of which destroys the reading frame in the legumin β subunit. A complete sequence of this gene shows considerable homology to leg A, not only in coding sequence, but also in the intron regions. Regions of 5' flanking sequence are also homologous, but the 3' flanking sequence is non-homologous, although it does contain two polyadenylation signals. It seems likely that leg A and leg D originally worked as a tandem pair of genes (having originated by gene duplication) but leg D was 'silenced', possibly as a result of change to its coding sequence.

REFERENCES

1. Lycett, G.W., Croy, R.R.D., Shirsat, A.H. and Boulter, D. 1984. The complete nucleotide sequence of a legumin gene from pea (Pisum sativum L.) Nucleic Acids Research. 12, 4493-4506.

THE ISOLATION, ANALYSIS AND EXPRESSION OF SPECIFIC PEA SEED, STORAGE PROTEIN GENES.

R. CASEY, C. DOMONEY, T.H.N. ELLIS AND J. CASTLETON.

The project is to identify related storage protein genes which are differentially expressed in developing pea seeds and to establish the molecular basis of such differences. The genes selected for such studies are those that code for the various forms of legumin. The first step is to characterize various cloned legumin cDNAs in terms of cross-homology and hybridization patterns to genomic DNA sequences, and to measure the relative amounts of the mRNAs corresponding to these cDNAs to be found in developing seeds.

(a) cDNA studies

Three cDNA plasmids (pCD32, pCD40 and pCD43) have been identified as corresponding to legumin-related mRNAs (ref. 1 and unpublished). In hybrid-release/translation experiments each clone selects a mRNA population which on cell-free translation produces a distinct set of polypeptides. pCD32 selects predominantly mRNA for an ~80,000*Mr* polypeptide and small amounts of mRNA for polypeptides in the size range ~63-65,000*Mr*; pCD40 selects predominantly mRNA for the ~63-65,000*Mr* polypeptides; and pCD43 selects predominantly mRNA for polypeptides of ~60,000*Mr*.

Southern blot analysis of the cross-hybridization behaviour of the three cloned cDNAs under different stringencies of hybridization indicate that pCD43 is distinct from pCD40 and pCD32, while the two latter share some homology.

Experiments are in progress to estimate the relative steady-state concentrations of the mRNAs corresponding to these cDNAs in pea seeds at a number of different stages of development.

Hybridization of cDNAs to pea genomic DNA

Hybridization of the cDNA inserts from pCD32, pCD40 and pCD43 to Southern blots of restricted genomic pea DNA (fractionated on agarose gels) shows that each cDNA hybridizes to a distinct set of restriction fragments. The pattern obtained with pCD43 is similar to that presented by Croy et al. (ref. 2) as that for a major legumin cDNA. pCD40 and pCD32 are presumably from mRNAs corresponding to other legumin-related genes. There is some variation in the patterns of hybridization betwen different pea genotypes.

(b) Isolation of legumin genes from genomic libraries

Pea genomic libraries have been constructed in bacteriophage λ and screened with inserts from pCD32, pCD40 and pCD43.

(c) Protein studies

A minor legumin has been purified. It represents only a few percent of the total legumin and differs from the main part of the legumin in having α-polypeptides which are apparently appreciably larger and β-polypeptides which are apparently slightly smaller than the majority. It is not known at present how the protein relates to pCD32, pCD40 and pCD43.

REFERENCES

1. Domoney, C. and Casey, R. (1984) Storage protein precursor polypeptides in cotyledons of Pisum sativum L. Identification of, and isolation of a cDNA clone for, an 80,000-Mrlegumin-related polypeptide. Eur. J. Biochem. 139, 321-327.
2. Croy, R.R.D., Lycett, G.W., Gatehouse, J.A. Yarwood, J.N. and Boulter, D. (1982) Cloning and analysis of cDNAs encoding plant storage protein precursors. Nature 295, 76-79.

ISOLATION AND CHARACTERIZATION OF CEREAL SEED PROTEIN GENES.

M. KREIS, B.G. FORDE, P.R. SHEWRY, M. WILLIAMSON, J. PYWELL, H. LEWIS, A. HEYWORTH and B.J. MIFLIN.
Rothamsted Experimental Station, Harpenden, Herts. U.K.

Prolamins are a group of alcohol soluble storage proteins found only in the seeds of the gramineae. The prolamins of barley (Hordein), Rye (secalins) and wheat (gliadins) account for 50% of the mature seed proteins and because of their deficiency in lysine and other essential amino acids they determine the nutritional quality of the grain. These proteins are tissue-specific and their synthesis appears to be temporarily regulated during endosperm development. Using poly A^+ RNA from developing endosperms of barley Rye and wheat, we have constructed cDNA libraries and isolated and characterized cDNA clones coding for the sulphur-rich (B_1 and B_3 hordeins, γ secalins), the sulphur poor (C hordeins, W secalins) and high molecular weight (HMW and D hordeins) storage proteins. Each group of storage proteins is encoded by a multigene family. The B_1 and B_3 hordein clones were used to probe a barley genomic library. A B hordein genomic clone has been characterized and the nucleotide sequence determined. The clone contains a gene coding for a B1 hordein of 274 amino acids. The sequence confirms a partial cDNA sequence. The signal sequence of 19 amino acids is homologous to that reported for α-gliadin and to that of zein (although there is little other evidence of homology between hordeins and zeins). The coding sequence shows two different structural domains. The first domain of about 80 amino acids is very rich in glutamine and proline and is made up of a short repeated sequence (consensus: PQQPFPQQ). The second domain of 194 amino acids has a non-repeated structure which is poorer in proline but contains about 4% cysteine. These domains are very similar in structure and composition and have a similar sequence to those determined for the sulphur-rich prolamins of wheat and rye. The barley genomic clone contains no introns. The promotor region contains the sequence 5'TATAAATA'3 between -79 and -

72, and two potential cat boxes 5'CCAAACATCCA3' between -153 and -143 and 5'CCAAAAATCCA3' between -167 and -157 (numbered relative to the translation initiation codon). These sequences match consensus sequences found in other genes. Potential polyadenylation signals are located about 60, 120 and 130 nucleotides downstream from the translation termination codon. Because of the repeated structure and the restricted distribution of the prolamins it has been suggested that they are of recent evolutionary origin. Our results suggest that this is only partially correct and that the non-repetitive domains of the sulphur-rich prolamins of the Triticeae (wheat, barley and rye) are homologous to a family of enzyme inhibitors (i.e. Trypsin inhibitor, α-aylase inhibitor) and storage proteins present in cereals, castor bean (Ricinus communis L) and oilseed rape (Brassica napus L.). We conclude that the prolamins have their origin in an old gene family present before monocotyledons and dicotyledons diverged. The barley salt-soluble protein fraction contains several tissue-specific major proteins (e.g. Chymotrypsin inhibitors, β-amylase and protein Z) that have been well characterized and which are rich in lysine and behave as storage proteins in many respects. Using poly A^+ RNA from barley (Hiproly) we have constructed cDNA libraries. A synthetic 14-bases long oligo nucleotide was used to probe the libraries for chymotrypsin inhibitor related DNA sequences. A chymotrypsin inhibitor clone has been identified, characterised and the nucleotide sequence determined. Both the B hordeins and the chymotrypsin inhibitor sequences will be used to study the tissue specific developmental regulation of expression of these genes.

CLONING AND MOLECULAR ANALYSIS OF GENES FOR WHEAT ENDOSPERM PROTEINS

R.D. THOMPSON, D. BARTELS, N.P. HARBERD and R.B. FLAVELL

The major wheat storage proteins, gliadins and glutenins represent two complex multigene families of nutritional and technological importance (1,2). An investigation into the structure and organisation of these genes has been initiated by the identification of cDNA clones coding for α/β, γ and ω-gliadins, and for the high molecular weight (HMW) subunits of glutenin.

The cDNA clones have been characterized in terms of the polypeptides specified by related mRNAs, on 1D and 2D gel systems and by DNA sequencing. DNA sequences of ω-gliadin and HMW glutenin coding regions reveal the presence of long stretches of short in-frame direct repeats, which have implications for the evolution of these genes.

In order to determine the chromosomal location and copy number of homologous sequences in the wheat genome, cDNA clones from different gliadin and glutenin sub-families were hybridized to endonuclease-digested wheat DNA. The α/β gliadin sequences are present at 25-35 copies per haploid genome on chromosomes 6A, 6B and 6D, and one group of γ-gliadin sequences is present at 10-15 copies/haploid genome on 1A, 1B and 1D. In the variety 'Chinese Spring', HMW glutenin genes are present at 5-10 copies/haploid genome with 2-3 genes on the long arms of chromosomes 1A, 1B and 1D.

Using lines lacking a single chromosome (nullisomic-tetrasomic series) we have shown that the HMW glutenin genes on chromosomes 1A in Chinese Spring are inactive, i.e. do not direct the synthesis of a detectable mRNA or polypeptide (3).

A HMW glutenin cDNA clone has been used as a probe for the

identification of cloned HMW glutenin genes in a bacteriophage lambda recombinant library. Clones have been obtained derived from the 1D and 1A chromosomes of Chinese Spring and these clones are being investigated by restriction enzyme mapping and DNA sequencing.

REFERENCES

1. Kasarda, D.D. (1976). Wheat proteins. In Pomeranz, Y. (ed.) Advances in Cereal Science and Technology Vol. 1, St. Paul Minnesota, American Association of Cereal Chemists, p. 158-236.
2. Payne, P.I., Holt, L.M., Jackson, E.A. and C.N. Law (1984). Wheat storage proteins: their genetics and their potential for manipulation by plant breeding. Phil. Trans. Roy. Soc. London 304 p. 359-371.
3. Thompson, R.D., Bartels, D., Harberd, N.P. and R.B. Flavell (1983). Characterization of the multigene family coding for HMW glutenin subunits in wheat using cDNA clones. Theor. Appl. Genet. 67, p. 87-96.

IMPROVEMENT OF PROTEIN QUALITY IN BARLEY BY MEANS OF GENETIC ENGINEERING

S.K. RASMUSSEN, H.E. HOPP, H. GIESE, A. BRANDT and D. von WETTSTEIN
Department of Physiology, Carlsberg Laboratory
Gamle Carlsberg Vej 10, DK-2500 Copenhagen Valby, Denmark

Hordein, the alcohol-soluble lysine-deficient storage protein of the barley endosperm, is a complex mixture of polypeptides encoded by at least three different loci on chromosome 5. One codes for the B-hordein polypeptides with molecular weights from 27 to 38 kd. Protein Z, a lysine rich endosperm albumin with a molecular weight of 40 kd is coded for by a locus on chromosome 4. cDNAs encoding hordein and protein Z polypeptides have been constructed and characterized in order to provide probes with which to investigate possible ways of changing the nutritional quality of the barley grain.

Ten different cDNA clones encoding B-hordein polypeptides have been identified. These can be divided into four groups, on the basis of their overall nucleotide sequence. A large number of nucleotide differences resulting in amino acid substitutions are found among and within the groups. Three of the groups have identical 3´non-coding regions, whereas the fourth group has a 25 nucleotide long deletion, and several nucleotide substitutions, in this region. The cDNA encoding protein Z covers the 180 carboxyterminal amino acids of a protein Z polypeptide. The 3´non-coding region is 130 nucleotides long with two putative polyadenylation signals. This region is followed by the poly-A tail. Hybridization of B -hordein and protein Z cDNAs to restriction endonuclease fragments of barley nuclear DNA suggests that the B-hordein polypeptides are coded for by a multigene family with 10-15 members whereas protein Z is probably coded for by a single gene. Abundant mRNA sequences ranging in size from 1200 to

1400 nucleotides were detected by hybridization to the B-hordein probes, whereas the protein Z mRNA was found to be 1800 nucleotides long. These sizes are sufficient to code for the B hordein and protein Z precursor polypeptides, respectively.

The amount of mRNA encoding B-hordein and protein Z polypeptides was analysed in developing endosperms, from spikes cultured in liquid media containing different amounts of ammonium nitrate. Increasing amounts of mRNA were synthezised in response to increasing nitrogen application. The amount of protein Z mRNA increased liniarly with increasing nitrogen in the medium. In contrast, stimulation of transcription of the B-hordein genes reached saturation at the lowest nitrogen concentrations. Hordein and protein Z mRNA levels are effectively increased by application of nitrogen after the onset of storage protein synthesis. Removal of nitrogen in the medium during storage protein synthesis leads to a short term decrease in the amount of hordein and protein Z mRNAs. It is thus possible to modulate the levels of these mRNAs by changing the nitrogen supply to the spike.

REFERENCES

1. S.K. Rasmussen, H.E. Hopp, A. Brandt, I. Svendsen and J. Hejgaard: A cDNA clone for protein Z, a major barley endosperm albumin. Carlsberg Res. Commun. 49, 385-390, 1984.
2. H. Giese and H.E. Hopp: Influence of nitrogen nutrition on the amount of hordein, protein Z and β-amylase messenger RNA in developing endosperms of barley. Carlsberg Res. Commun. 49, 365-383, 1984.

CLONING AND SEQUENCING OF THE HOR-2 LOCUS IN BARLEY

V. CAMERON-MILLS, A. MOMTEMBAULT and A. BRANDT
Department of Physiology, Carlsberg Laboratory
Gamle Carlsberg Vej 10, DK-2500 Copenhagen Valby, Denmark

The B-hordein storage polypeptides of the barley endosperm are coded for by a locus on the barley chromosome 5. Southern blot analysis of restriction endonuclease fragments of barley DNA, using cDNA encoding B-hordein polypeptides as a probe, suggests that the locus is complex and composed of at least ten copies of B-hordein genes. In order to elucidate the molecular structure of this gene cluster we have constructed a library of barley genomic DNA in the bacteriophage lambda vector L 47.1. Following screening of the library with a B-hordein cDNA probe one recombinant clone containing a 22 kilobase pair barley DNA fragment was selected for further analysis by restriction mapping and nucleotide sequencing. One complete B-hordein structural gene is contained within a 2.9 kilobase pair Hind III fragment in this clone. By cloning overlapping fragments of barley DNA to the isolated B-hordein genomic clone we aim to characterize a larger portion of the B-hordein gene cluster.

CLONING AND SEQUENCING OF A CARBOXYPEPTIDASE GENE FROM BARLEY

G.B. FINCHER, A.J.S. CHOJECKI and A. BRANDT
Department of Physiology, Carlsberg Laboratory
Gamle Carlsbergvej 10, DK-2500 Copenhagen Valby, Denmark.

Malt carboxypeptidase (CPD-M) is an exopeptidase involved in protein degradation in germinating barley grains. The enzyme consists of two subunits, each made up of two different polypeptides (molecular weight 31,000 and 19,000) whose amino acid sequence is known in part. In-vitro translation of mRNA isolated from scutella at 2, 5 and 8 days after germination, followed by immunoprecipitation with CPD-M antibodies, revealed a marked increase in the amount of mRNA coding for the CPD-M large polypeptide between 2 and 8 days. CPD-M mRNA was partially purified by sucrose gradient fractionation. Double-stranded cDNA was synthesised, inserted by G-C tailing into the Pst 1 site of pBR 325 and cloned by transformation of E.coli LE 392 cells. By differential hybridisation to 2, 5 and 8 day mRNA, clones were selected that are likely to contain CPD-M coding sequences. Following subcloning in M 13 mp8, the nucleotide sequence of these cDNAs is being determined using the dideoxy-chain-termination method.

STRATEGIES FOR CLONING ZEIN REGULATORY GENES IN MAIZE

R. MAROTTA*, M. MOTTO, N. DI FONZO, M. BREMENKAMP, C. SOAVE* and F. SALAMINI
Istituto Sperimentale per la Cerealicoltura, Via Stezzano 24, 24100 BERGAMO, Italy
*Istituto Biosintesi Vegetali, C.N.R., Via Bassini 15, 20133 MILANO, Italy

A suitable model system for studying gene regulation in higher plants is represented by the Opaque-2 and Opaque-6 loci of maize. Recessive mutations at both loci result in decreased expression, during development, of several unlinked zein genes, and to the disappearence of a 32kd protein (b-32), whose structural gene is Opaque-6. b-32 appears to play a key role in sustaining the coordinate expression of storage protein genes in maize endosperm, while Opaque-2 is postulated as a regulatory gene for Opaque-6 expression. A more detailed characterization of the interaction between Opaque-2 and Opaque-6 in regulating zein genes can result from the availability of molecular probes for these loci.

The identification of the gene product of Opaque-6 allowed us to select, out of a cDNA library made from maize endosperm polyA+ RNA, several clones complementary to b-32 mRNA as judged by hybrid-released translation followed by immunoprecipitation of translation products with purified b-32 antibody. A recombinant plasmid (pBC36) containing a cDNA insert of 850 bp has been choosen to probe the organization of Opaque-6 in wildtype maize strains and several mutants, and to study the expression of the locus at the RNA level. Preliminary results will be reported.

As far as the Opaque-2 locus is concerned, a similar cloning strategy couldn't be pursued, since the nature of its molecular product is not known. The ability of maize controlling elements, some of which have been biochemically characterized, to induce insertion mutants at a variety of loci, has suggested the alternative approach of cloning Opaque-2 sequences by means of physical linkage to an inserted element. The Ac-Ds controlling element system, for which molecular probes are available, has been used to induce mutations at the Opaque-2 locus. Four independent mutants have been isolated: these are somatically unstable, and the instability is under the control of the Ac element, thus behaving as Ds insertion mutations. Three of these (o2m114-1, o2m114-3 and o2m8-1) show a normal endosperm phenotype with sectors of opaque tissue. A fourth mutant (o2m114-2) has a recessive phenotype, and reverts somatically to wildtype in the presence of Ac. During this year, biochemical studies have been concentrated on this latter mutant, whose genetic behaviour is similar to other Ds insertion mutants confirmed at the molecular level. Efforts have been made in order to identify, among the many restriction fragments in the maize genome hybridizing with a Ds probe, a DNA sequence cosegregating with the o2m114-2 allele, in appropriate genetic materials: this approach however has not been successful, possibly due to the very complex pattern of Ds hybridizing fragments, and/or to limited homology between the Ds element present at Opaque-2 and the one used as the probe. Biochemical studies at the RNA level looked more promising: a 2 kb RNA species hybridizing with the Ds probe is present in the endosperm of maize stocks homozygous for the o2m114-2 allele, while no Ds transcription has been detected in the endosperm of strains carrying other Opaque-2 alleles. This result suggests that the Ds element inserted at Opaque-2 is cotranscribed with the locus, as already observed for a Ds induced mutation at the Adh-1 locus (M. Freeling, P. Starlinger, personal communication), thus indicating a possibility for cloning Opaque-2 sequences.

TRANSCRIPTION AND EXPRESSION OF ZEIN SEQUENCES IN YEAST

A. VIOTTI[1], E. MARTEGANI[2], C. COMPAGNO[2], I. CORAGGIO[1], B.M. RANZI[2], L. ALBERGHINA[2].
1) IST. BIOSINTESI VEGETALI - CNR - MILANO - ITALY.
2) DIP. FISIOLOGIA E BIOCHIMICA GENERALI - SEZ. BIOCHIMICA COMPARATA - UNIVERSITA' DI MILANO - MILANO - ITALY.

Recombinant plasmids from interspecific (E.coli-S.cerevisiae) vectors and a genomic maize fragment containing a zein gene have been constructed to study the transcription and expression of zein sequences under homologous or inducible heterologous promoters in yeast.
The 2000 bp Sal-Sal1 fragment of the zein gene E 19 (1) and its flanking sequences has been inserted in both YRp7 and pLGSD5 vectors (2,3) obtaining respectively Y7z and pL15. A third recombinant (pL26) under the yeast GAL10-inducible promoter has been obtained. Briefly, the BamH1-Xba1 fragment of 613 bp from the subclone pUE26 in pUC8 plasmid has been cut with SmaI and Hinc II enzymes, and inserted downstream the GAL promoter in the BamH1 site of pLGSD5 after Klenow polish by blunt end ligation. The frame of the zein sequence to the iso-1-cytochrome C sequence of pLGSD5 at the junction site was assessed by sequencing both strands and has the following codons: ATG.ACC.GGG.GGG.ATC. → zein, with the third and fourth codons derived from the Klenow treatment. Transformed yeast cells were grown in suitable medium in presence or absence of galactose to follow endogenous or induced transcriptions of zein genes. Fig. 1 reports the Northern blot analysis of the zein transcripts obtained in different growth conditions. A large amount of zein transcripts of proper size is clearly visible in Gal-induced cells transformed by pL26. Zein transcripts are also visible (in a longer exposure to the film) in yeast cell transformed with Y7z and pL15, in respect to own control, indicating that the zein promoter is recognized and that transcription is accurately initiated. Immuno electroblotting and dot immuno assays of total or ethanol

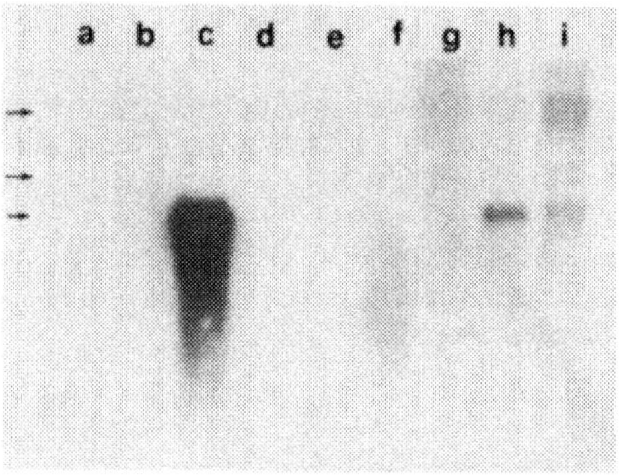

FIGURE 1. Northern blot analysis of RNA extracts from S.cerevisiae transformed with different recombinant vectors. Cells were broken with alumina phenol extracted twice and then the RNA selectively precipitated by the LiCl method. About 20 ug of RNA were loaded on each lane and fractionated by electrophoresis on a 1.5% agarose formaldehyde gel. The RNA was blotted on a nitrocellulose filter and the filter hybridized to Sal-Sal E19 fragment labelled by nick translation. Yeast strain X-4004 was transformed by: a) YRp7; b) Y7z; e) pLGSD5; f) and g) pL15; h) and i) pL26. Cells were grown in glucose (a,b,e,g,i) or galactose (f,h). Total RNA from maize endosperm (c) and from maize leaves (d) were also loaded as control. On a side, from top to bottom, arrow heads indicate the relative mobility of the two yeast rRNAs and zein mRNA.

soluble proteins reveal the presence of a zein polypeptide of expected mol. wt., suggesting that also transcription-termination and translation are properly carried out.

REFERENCES.

1. A. Spena, A. Viotti, V. Pirrotta 1983. Two adjacent genomic zein sequences: Structure organization and tissue specific restriction pattern. J. Mol. Biol. 169: 799-811.
2. K. Struhl, D.T. Stinchcomb, S. Scherer, and R.W. Davis 1979. High-frequency transformation of yeast: Autonomous replication of hybrid DNA molecules. Proc. Natl. Acad. Sci. USA 76: 1035-1039.
3. L. Guarente, R.R. Yocum, and P. Gifford 1982. A GAL10-CYC1 hybrid yeast promoter identifies the GAL4 regulatory region as an upstream site. Proc. Natl. Acad. Sci. USA 79: 7410-7414.

OUTLINE OF METHODOLOGIES FOR THE INDUCTION OF INSERTION MUTANTS AT MAIZE LOCI INVOLVED IN STARCH SYNTHESIS AND FOR THE ISOLATION OF THE CORRESPONDING GENE SEQUENCES

L. DEL GIUDICE
Istituto Internazionale di Genetica e Biofisica / CNR, Via G. Marconi 10, I-80125-NAPOLI

Du and Ae loci of maize

Among the various mutations affecting the percentage of amylose in maize starch[1], dull (du)[2] and amylose-extender (ae)[3] alleles enhance the sinthesis of amylose. The last mutation induces a strong increase of the amylose in the endosperm, and it has been widely utilized in corn breeding.

While other mutations affecting starch synthesis in maize have been biochemically characterized and cloned (wx and sh alleles)[4] ae and du alleles have not been. Biosynthetic routes leading to starch accumulation in maize endosperm have been frequently studied and critically discussed[1]. However, for this and other plants no definitive data are available on the enzymatic reaction specifically involved in amylopectin and amylose syntheses.

"Mutator" (Mu)

Mutator in maize was first described by Robertson[5]. It consist of a transposable elements transmitted in a non-mendelian fashion (90% of outcross plants transmit Mutator) and cannot be mapped on chromosomes; however, it is not a cytoplasmic genetic element[6]. The frequency of mutation in the presence of "Mutator" increases up to thirty-fold[7]. Since most of the newly-induced mutations are autonomously unstable[6], it is reasonable to believe that the autonomous transposable element inserts into maize loci generating somatic instability. At the molecular level, a DNA insert has been found in Mu induced mutations at the Adhl and Shl loci (reviewed in Freeling, 1984[8-9]).

The Mu visited locus can be cloned by using the transposable element as probe[10,11].

Description of the research work proposed

The research relates to the generations of new, Mu-induced mutations at the Ae and Du maize loci. The final goal of the research project is to test whether the Mu element is suited for tagging such selected loci of maize. If this will prove true, molecular clones from this loci could be obtained.

Wildtype plants carrying Mu will be pollinated with ae and du recessive pollen. New mutations will be revealed by the appearance of recessive phenotypes: those induced by Mu generally occur as cluster of mutated kernels and are frequently unstable. Unstable phenotypes will be selected after selfing plants deriving from the mutated kernels, to distinguish new and old recessive gametes.

Molecular characterization of the Mu induced mutations

DNA inserts have been found in Mu-induced mutations at the Adh1 locus in maize. Mu clones are available. They will be used as probes to analize the restriction pattern of the DNA from F2 (or backcross) plants segregating the new mutations. Since several copies of Mu are expected to be found, we shall investigate whether a particular "Mu-hybridizing" fragment cosegregates with unstable ae and du phenotypes.

Identification of Ae and Du clones

DNA will be purified from ae and du plants carrying a Mu insert at these loci. A genomic library will be constituted and first screened with a Mu probe. The "Mu-Library" will be differentially hybridized to total endosperm cDNA, made alternatively from wildtype or ae and du genotypes because most transposable elements induced mutations are due to a defect in transcription caused by insertion[12]. Flanking sequences corresponding to Ac and Du clones will be finally identified. This last part of the project will be facilitated having available other unstable mutants induced by transposable elements different from Mu such as Ac-Du or Spm.

References

1. Creech R.G., 1968 - Adv. Agron. 20: 275.
2. Cameron J., 1947 - Genetics 32: 459.
3. Vinegard M.L., 1952 - Maize Genet. Coop. News Lett. 26: 5.
4. Geiser M., 1982 - EMBO J. 1: 1455.
5. Robertson D.S., 1978 - Mutat. Res. 51: 21.
6. Robertson D.S., 1981 - Mut. Res. 84: 283.
7. Robertson D.S., 1980 - Genetics 94: 969.
8. Freeling M., 1984- Ann. Rev. Plant Phys. 35: 277.
9. Peterson P., 1978 - In Maize Breeding and Genetics, Walden Ed. p. 601
10. Birgham P.M., 1980 - Genetics 95: 341.
11. Searles L.L., 1982 - Cell 31: 585.
12. Scherer S., C. Mann and R.W. Davis, 1982 - Nature 298: 815.

The genetics of secondary metabolite production in higer plants: the flavonoid genes of Petunia as a model system.

J.N.M. Mol[1], R. Koes[1], E.A. van den Berg[1], H.J. Reif[2], F. Kreuzaler[2], E. Veltkamp[1].

[1]Dept. of Genetics, Free University, De Boelelaan 1087, 1081 HV Amsterdam

[2]Max-Planck-Institut für Züchtungsforschung, Egelspfad, D-5000 Köln 30

Many plant species produce a class of compounds called secondary metabolites which have been exploited successfully in agriculture, industry and pharmacy (e.g. flavors, antibiotics, drugs, hormones, alkaloids etc.). When controlled production of such substances is aimed at, one needs to know all individual steps of the biosynthetic pathway, the genes and gene products involved and their regulation. To date such information on economically-important secondary metabolite pathways is not available yet.
The flavonoid gene system of Petunia hybrida constitutes an elegant modelsystem that enables us to study regulatory aspects of secondary metabolite production for the following reasons:

- the majority of the genes involved in the flavonoid pathway of Petunia are identified and characterized
- the activity of most of the gene products can be measured
- regulation by light hormones and transposable elements has been reported
- flavonoid genes are coordinately expressed and tissue-specifically expressed
- mutants are available in all steps of the biosynthetic pathway

We have begun cloning genes for chalcone synthase (CHS) and glucosyltransferase (3GT) plus flanking sequences (cf. fig. 1, simplified pathway)

We constructed a genomic library of Petunia strain V30 (which has all the colour genes dominantly expressed) in the Eco RI site of λ GT wes and screened this library with CHS cDNA. Five different clones were identified containing inserts ranging in size from 2,6 to 6,8 kb. Fig. 2 shows the result of a recloning experiment in plasmid pUC18:

One way to study the problem of (tissue-specific) gene expression is to reintroduce wildtype or mutagenized genes in plant mutants. To be meaningful however, one needs to have available the in vivo expressed copy of the gene. Since our initial cloning experiments suggest the presence of multiple copies of the CHS gene in strain V30, we are in the process of

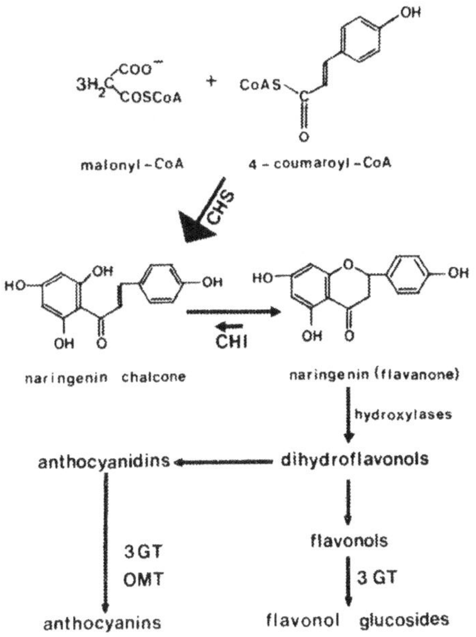

Fig. 1.

Simplified scheme of the pathway of flavonoid biosynthesis in P. hybrida.

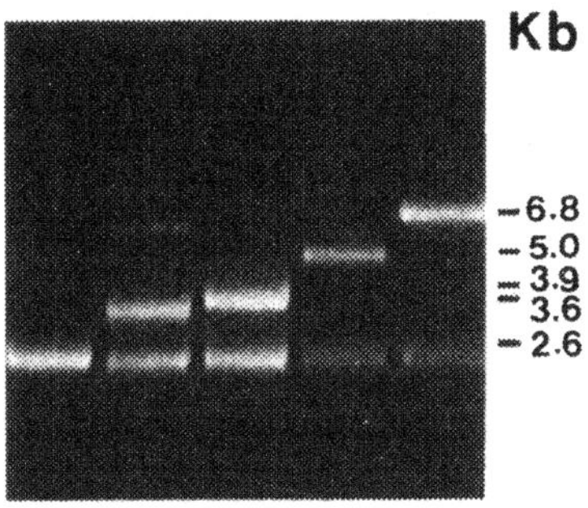

Fig. 2.

Agarose gel electrophoresis of CHS clones in plasmid pUC18x Eco RI.

identifying V30 cDNA clones that match the different chromosomal inserts. Such an analysis allows one to identify strong and weak CHS promoters. Such elements will be used to transform a conditional CHS mutant of Petunia recently described[1]. Another gene system currently under investigation is the 3GT locus which codes for a 3' glucosylating activity. In analogy to Zea mays Bz we suspect it to be under transposition control in certain unstable mutants of Petunia. Currently we are trying to obtain evidence for this hypothesis by hybridizing Petunia genomic DNA with Bz DNA (kindly donated by Nina Fedoroff, Washington) and with cloned DNA from a variety of mono- and dicotyledonous transposable elements.

Reference

Mol et al. (1983) Regulation of flavonoid gene expression in Petunia hybrida: Description and partial characterization of a conditional mutant in chalchone synthase gene expression. Mol. Gen. Genet. 192: 424-429.

CLONING OF A LYSINE DECARBOXYLASE GENE FOR THE GENETIC TRANSFORMATION OF LUPINUS PLANTS.

L. FECKER, H. BEIER AND J. BERLIN

GBF-GESELLSCHAFT FÜR BIOTECHNOLOGISCHE FORSCHUNG MBH,
D-3300 BRAUNSCHWEIG, FRG.

Our group has recently shown that inducible enzymes connecti primary and secondary pathways seem to be decisive for the initiation of secondary metabolism in plant cell cultures. Such enzymes are our first targets in our investigations of manipula ing secondary metabolism genetically. In one of our projects we want to transform Lupinus cells with the gene of a bacterial lysine decarboxylase (LDC) as plant LDC seems to be a bottlenec of quinolizidine alkaloid biosynthesis in Lupinus cultures. In the progress report 1983 (1) we have described the purification and characterization of LDC from Hafnia alvei, the preparation and purificaton of LDC antibodies, and the first detection of a LDC positive clone (462) from our gene bank constructed in E. coli HB101. Here we report on the progress in cloning the gene and characterizing the enzyme.

The LDC-positive clone 462 carried an insert of about 7.5 kB As LDC subunits have a MW of 75.000 the coding capacity of the insertion of pLD 462 was about 3 times too big. To reduce the size of the inserted DNA pLD 462 was partially digested with Sau 3a. DNA fragments with 1,5-2,5 and 2,5-5 kB were isolated and ligated into the BamH1 site of pBR322. HB101 was transformed with the recombinant plasmids as before (1). Two positive clones 1-101 and 2-19 were screened from this new gene bank. The plasmids pLD1-101 and pLD2-19 carrried inserts of 3.3 and 2.9 kB. We assume that the LDC gene has a size of 2-2.3 kB. No positive clones were found in the experiments where DNA fragments of pLD462 below 1.5 KB were used. For DNA sequence analysis pLD1-101 and pLD2-19 were mapped with several restriction enzymes. The restriction map of that part of the insert which is common to all LDC positive clones (462 and subclones 1-101 and 2-19) is given in the Figure.

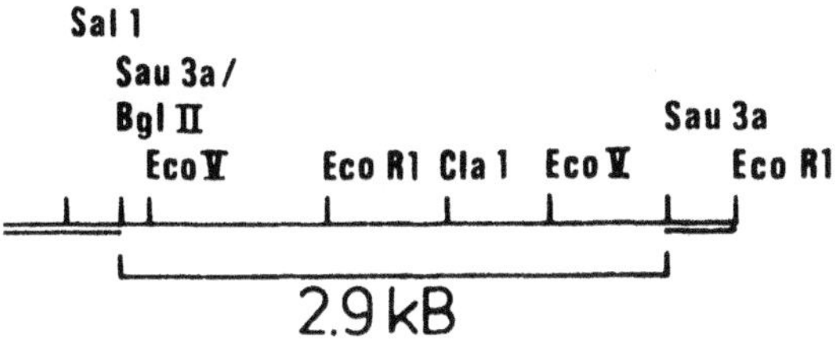

Restriction map of the 2.9 kB insert of pLD2-19

From the first gene bank two positive clones (462 and 1820) were isolated according to the applied ^{125}J-filter immunoassay (1). While a sonicate of 462 showed high LDC activity, the sonicate of 1820 contained no LDC activity. The precipitated protein of 1820 had indeed a lower MW than LDC when subjected to SDS PAGE. The plasmid pLD 1820 will be further characterized to describe which part of the LDC gene has been lost. The LDC synthesized by clone 462 was found to be identical to purified LDC of Hafnia alvei according to SDS PAGE. Minicells containing the plasmid of clone 462 synthesized a protein that has the same mobility as LDC in SDS PAGE.

Before the cloned gene can be used for transformation of plant cells it is necessary to localize the start of the gene.To aid the DNA sequence analysis, the amino acid sequence of the NH_2-terminus of purified LDC was determined up to the 12th amino acid. The following sequence was determined.

? - Ala - Ile - Asp - ? - Lys - Phe - Gly - (Tyr) - Asp - Ala -Arg

An oligonucleotid mix will be chemically synthesized and used as probe for detecting a homologous sequence which should be the start of the gene.

Acknowledgement: We are grateful to Dr. Hoppe who sequenzed the protein for us.

References:

J. Berlin, H. Beier, L. Fecker, W. Noé, H.U. Schairer; Progress Report 1983, Biomolecular Engineering Programme of CEC, Brussels pp. 126-127.

NUCLEAR GENES EXPRESSED DURING THE DEVELOPMENT OF YOUNG BARLEY LEAF CELLS

R.BARKARDOTTIR, B.FRIIS JENSEN, J.D.KREIBERG AND K.GAUSING

We wish to study the changing patterns of nuclear gene expression during leaf cell development and are particularly aiming at the controls of chloroplast development. As developmental system we have chosen the first leaf of 7 day old barley plants because in graminaceous plants is cell age proportional to the distance from the leaf base. We have determined the growth characteristics of our barley variety, Bomi, and analyses of basic biochemical parameters in sections of the leaf are in progress such that gene expression studies can be correlated with macromolecular synthesis on a whole leaf basis.

A major part of the programme of course includes the isolation and analysis of nuclear chloroplast genes but also genes for cytoplasmic macromolecules will be studied since the development of chloroplasts cannot be seen isolated from the development of the leaf cells as a whole. The source of genes is an ordered cDNA library (4800 clones) constructed from polyA mRNA from the basal 4 cm of 7 day old barley leaves. About 300 clones contain sequences from the large cytoplasmic rRNA habitually part of cDNA libraries. The remaining 4500 clones were subdivided by hybridization with cDNA from green leaves, etiolated leaves and roots into groups representing high, medium and low level classes of polyA mRNA as well as green leaf specific, leaf specific or "constitutive" classes. All clones representing the more abundant types of polyA mRNA have been grouped by cross colony hybridizations and hybridization with the genes for chlorophyll a/b binding protein (from Pea) and the small subunit of RuPCase (from wheat). One group of clones were identified by hybridization to a porcine gene coding for the nuclear/cytoplasmic double-function protein, Ubiquitin. The clone was kindly donated by A. Wind and J. Vuust in our department. The tally of genes isolated from the cDNA library is summarized in Table 1. The similarity of the two unknown types of clones, pKG1761 and 2101, to the clones for chlorophyll a/b binding protein and pSS, respectively, indicate that they also code for chloroplast proteins, a notion which we hope to confirm by further studies.

Table 1

Gene	No of clones	Expression
Chlorophyll a/b binding protein	33	Green leaf specific
Unknown type pKG1761	26	Green leaf specific
Small subunit of RuBPCase	19	(Green) leaf specific
Unknown type pKG2101	16	(Green) leaf specific
Unknown type pKG239	8	Constitutive
Unknown type pKG657	5	Constitutive
Ubiquitin	4	Constitutive
Chloroplast ribosomal protein L15	1	Not determined

The last clone listed in Table 1 deserves a special comment. It was originally isolated in a separate series of experiments based on weak homology detected between a DNA fragment from E. coli carrying two ribosomal protein genes (one of which is rplO) and cDNA from young barley leaf cells. The E. coli DNA fragment was used to enrich a polyA mRNA preparation about 500 fold for the L15 messenger by hybridization at low stringency. The selected polyA mRNA was converted into double-stranded DNA and cloned in pBR327. One of the four types of clones obtained was shown by DNA sequence determination to be homologous to the E. coli gene, rplO, coding for L15. The homology is 34% at the amino acid level (of the 2/3 of the gene found in the cDNA clones) and 40% homologous at the nucleotide level. Formally we have not proven that the cDNA clone codes for a chloroplast r-protein but it is unlikely that a cytoplasmic r-protein would show this degree of homology to an E. coli protein and the polyA mRNA for any mitochondrial r-protein would not be sufficiently abundant to be isolated by the procedure used. It will now be very interesting to arrive at the complete protein sequence, including the likely transit peptide as well as surrounding and possible intervening DNA sequences at the chromosomal locus.

THE LIGHT-DEPENDENT CONTROL OF GENE EXPRESSION IN BARLEY

H. GIESE[+], K. STEINMÜLLER[+], A. BATSCHAUER[+], E. MÖSINGER[*] and K. APEL[+]

[+] Botanisches Institut der Christian-Albrechts-Universität, Olshausenstrasse 40, D-2300 Kiel

[*] Biologisches Institut II der Universität, Schänzlestrasse 1, D-7800 Freiburg

The light-induced greening of etiolated barley plants is used as a model to study the light-dependent control of plastid development. Upon illumination a rapid transformation of etioplasts to chloroplasts is induced. The effect of illumination does not only include the light-dependent chlorophyll synthesis but also the appearance or decline of specific proteins within the plastid membrane fractions. So far two of these proteins have been studied in more detail: the light-harvesting chlorophyll a/b protein (LHCP) and the NADPH-protochlorophyllide oxidoreductase (PCR).

The levels of translatable mRNAS coding for these two proteins are affected by light. While the translatable mRNA level coding for the LHCP increases drastically upon illumination, the mRNA level encoding PCR rapidly decreases. This reciprocal effect of light is mediated by phytochrome P_{fr}.

Using cloned cDNAs as hybridization probes we have demonstrated in dot-blot experiments that the light-induced changes of the two translatable mRNAs are both paralleled by changes in the steady-state concentrations of the mRNA sequences.

Furthermore, the transcriptional rates of the two genes in barley leaves grown under different light regimes were quantitated by monitoring nascent nuclear RNA transcripts in pulse-labelled nuclei. Similar to the results on the light-induced changes of the steady-state concentrations of the two mRNA sequences the transcription rates of the corresponding genes are regulated reciprocally under the influence of phytochrome.

This inverse relationship between the light-dependent regulation of the PCR and the LHCP demonstrates that a single external light stimulus, mediated by phytochrome can both increase and decrease the transcription of certain genes in higher plants.

STUDIES OF CHLOROPLAST DNA FROM *HELIANTHUS* SPECIES : GOALS AND OBSTACLES

Y. MANSAIS, C. PEREZ, A. BERVILLE *Station d'Amélioration des Plantes*
J. MARTIN-TANGUY *Station de Physiopathologie*
I.N.R.A. DIJON France

Ph. HEIZMANN *Departement de Biologie Générale et Appliquée LYON*

A set of wild species of the genus *Helianthus* was chosen because they could lead to cytoplasmic male sterility when crossed by sunflower (1). The first results on the chloroplasts DNA (cpDNA) studies were reported. Some isogenic couples (male-sterile, maintainer) of the french cytoplasm (2) were compared after digestion of the cpDNA by endonucleases (5 enzymes). No difference was revealed while for one of the couples some differences were reported at the mitochondrial DNA level (3).

Several unsuccesses forthe hydrolysis were met depending upon the line or the stage of the plants. The younger the plant stage the clearer the cpDNA fraction and consequently the latter is easier to hydrolysis by endonuclease. These unsuccesses increase for the wild species even the great deal of ways taken to protect the cpDNA. So that few diagrams were obtained. Nevertheless a cpDNA variability was revealed that incites us to enlarge the field of the study.

In order to explain such a failure to obtain the cpDNA likely to be hydrolysed the contaminating compounds were elicited. Quinons and free phenolic compounds were present but by cleaning them off the cpDNA remains unhydrolysable. CpDNA purification by CsCl gradient, preparative electrophoresis or by phenol-chloroform extraction does not improved its digestibility.

Such a DNA after partial hydrolysis (HCl) releases amins (putrescine and tyramin) plus some phenolic compounds while whole hydrolysis releases large amounts of putrescine, spermidine, spermine, tyramine and noradrenaline.

These molecules tightly bound to the cpDNA could prevent the hydrolytic process.

Questions arise from these facts : the main is about the nature of the bound between these molecules and the cpDNA.

On an other hand ways are tried to overcome these obstacles and to extend the study of the cpDNA variability among the species of the genus *Helianthus*.

REFERENCES

1. SERIEYS H. 1984. Communication at IInd Meeting Eucarpia on Sunflower held in Leningrad. 16-21 April.
2. LECLERCQ P. 1969. Ann. Amél. Plantes, 19, 99-106.
3. LEROY P., BAZETOUX S., QUETIER F., DELBUT J., BERVILLE A. 1984. Submitted for publication.

REPEATED SEQUENCES AND MOLECULAR ORGANIZATION IN WHEAT MITOCHONDRIAL DNA

B. LEJEUNE, F. QUETIER, D. FALCONET, M.F. JUBIER, S. DELORME, M. SEVIGNAC and E. DELCHER.

Laboratoire de Biologie Moléculaire Végétale,LA CNRS n° 40, Université Paris-Sud, ORSAY, France.

Plant mitochondrial DNA contains repeated sequences (1, 2, 3 ; FALCONET et al., submitted). A thorough investigation ot their occurence in wheat mt DNA was made by using a complete library of this genome constructed by with pBR 322 and pHC 79. There are 52 different Sal I fragments whose molecular weight range from 0.4 kb to 29 kb. Each fragment has been obtained individually cloned and each clone was used as a radioactive probe in hybridization with Sal I restricted mt DNA. Cross-hybridization between different Sal I fragments allowed us to evidence ten kinds of repeated sequences, the size of which ranged from 0.3 to 6 kbp. Some of these sequences do not contain a Sal I site and are found each in four different Sal I fragments. An instance of this has been published (3) concerning the repeated sequence harbouring the 18 and 5 S r-RNA genes, where the presence of two different kinds of sequences on each side generates four fragments evoking a crossing-over. The same type of structure has also been evidenced for the case of the repeated sequence containing the 26 S r-RNA gene (with an internal Sal 1 site, FALCONET et al., submitted) and for two other repeated sequences.

From these results it can be inferred that recombinationnal process involving these repeated sequences occur in mt DNA (whether the whole length of the repeated sequence participates or whether there is only a site-specific recombination is still to be determined).

From the presence of several repeated sequences in a single genome, one may expect the occurence of several different kinds of molecules generated by a recombination-excision process. This explains the heterogeneity of the population of mt DNA molecules in higher plants (4). Structural instances of this have been published (1, 2). In the case of wheat, a similar situation has been found. A master molecule of 430 kbp can be drawn which includes all the original sequences and the repeated sequences : intramolecular recombination allows to generate different kinds of molecules.

REFERENCES

1. Palmer, J.D. and Shields, C.R. (1984) Nature 307, 437-440.
2. Lonsdale, D.M., Hodge, T.P., Fauron, C.M.-R. and Flavell, R.B. (1983) in ICN-UCLA Symposia on Molecular and Cellular Biology, Golberg, R.E. Ed., Volume 12, New Series, pp. 445-456, Alan R. Liss, New York.
3. Falconet, ., Lejeune, B., Quétier, F. and Gray, M.W.(1984), EMBO J., 3, 297-302.
4. Quétier, F. and Vedel F., (1977) Nature 268, 365-268.

WHEAT MITOCHONDRIAL DNA CONTAINS MULTIPLE ars-TYPE SEQUENCES

B. LEJEUNE, F. QUETIER, M.F. JUBIER and E. DELCHER.

Laboratoire de Biologie Moléculaire Végétale, LA CNRS n° 40, Université Paris-Sud, ORSAY, France.

The analysis of a recombinant cosmid library of wheat mitochondrial DNA has revealed the multimolecular organization of this genome (1). The occurence of at least ten sets of different repeated sequences can lead to various recombinationnal events which generate molecules of different sizes (circular or linear as well). The replication autonomy of such molecules is thus a key to their relative stoechiometry (2) ; the occurence of ars-type sequences has been studied by using a yeast system. The Hind III fragment carrying the URA 3 yeast gene has been inserted into pBR 322. Wheat mitochondrial DNA has been restricted by Eco RI and inserted into the corresponding site of the hybrid vector. The resulting plasmids were used to transform (ura -, his -, trp -) yeast cells and as much as 3,000 transformants per ug of DNA have been recovered. Plasmids were extracted from selected colonies and used to transform E. coli HB 101 cells. The amplified plasmids were shown to transform back ura - yeast cells.

The identification of the restriction fragment(s) carrying these ars -type sequences is achieved by nick-translation of each plasmid and by hybridization to the set of 52 individually cloned Sal I fragment which constitute the wheat mitochondrial DNA library. Several Sal I fragments (D, F, H, L, R, S, U for instance) carry ars sequences ; they are widely scattered along the 430 kbp master circular molecule and do not show peculiar location relative to the repeated sequences.

These results obtained on a mitochondrial DNA can be compared to the 2 ars sequences recently identified by the group of A.J. Kool on the chloroplastic DNA of Petunia hybrida (3). According to the results recently obtained by the group of J.D. ROCHAIX on the chloroplastic DNA of Chlamydomonas reinhardtii (4) it seems however that these ars-type sequences are not related to genuine ori sequences and that AT-rich stretches can induce DNA replication in yeast.

The study of the replication autonomy of the different molecules which constitute the wheat mitochondrial genome thus requires complementary approaches such as electron microscopic observation of mt DNA molecules isolated from psoralen-treated cells.

REFERENCES

1. Quêtier, F., Lejeune, B., Falconet, D. and Jubier, M.F. (1984),in Molecular Form and Function of the Plant Genome, Van Vloten-Doting ,L.,Groot, G. and Hall, T., Eds. Plenum Press, in press.
2. Quêtier F., Lejeune B., Delorme S., and Falconet D., (1984), Molecular organization and expression of the mitochondrial genome of higher plants. Encyclopedia of Plant Physiology, New Series vol. 18, Springer Verlag, in press.
3. Overbeeke, M.A., Haring, I., Hartingsveldt, W.V., De Haas, J.M. and Kool, A.J. (1984), in NATO-ASI-FEBS Advanced Course on Molecular Form and Function of the Plant Genome, van Vloten-Doting L, Groot, G. and Hall, T., Eds. p.B-21.
4. Rochaix, J.D., Vallet, J.M., Erickson, J., Spreitzer, R. and Goldschmidt-Clermont, M. (1984),in Molecular Form and Function of the Plant Genome, Van Vloten-Doting, L., Groot, G. and Hall, T., Eds. Plenum Press, in press.

HYBRIDIZATION STUDY OF THE PLASMID-LIKE MOLECULES FOUND IN MITOCHONDRIA OF VICIA FABA L.

J.P. GOBLET, M.C. FLAMAND and M. BRIQUET

Laboratoire d'Etude de l'Hérédité Cytoplasmique, Université de Louvain, Place Croix du Sud, 1, B-1348 Louvain-la-Neuve (Belgique).

1. INTRODUCTION

Many small supercoiled plasmid-like DNA molecules have been found within the mitochondria of different plants (Zea mays (1), faba bean (2, 3, 4), sugarbeet (5, 6)). Variations in the number and the size of these small supercoiled DNA molecules seem to be correlated to the cytoplasmic male sterility. But still little is known about their functions and their origins.

In order to better understand their relation with the cytoplasmic male sterility trait, we have studied the small plasmid-like DNA molecules found in the mitochondria of Vicia faba L. By electron microscopy and biochemical analysis, 2 small supercoiled DNA molecules of 1700 and 1420 base pairs were found in the mitochondria of the fertile and the male sterile 350 and 447 cytoplasms (2). A line containing the male sterile cytoplasm 350 was distinguished from the fertile cytoplasm and from the male sterile 447 cytoplasm by the presence of an additional supercoiled DNA molecule of 1540 bp.

All the mitochondrial plasmid-like DNA molecules have been cloned in bacterial plasmids and have been used in hybridization study against mitochondrial, nuclear and chloroplastic DNA.

2. RESULTS

With the cloning of the mitochondrial plasmid-like DNA molecules of Vicia faba L., we have showed the existence of a second supercoiled DNA molecule of 1700 bp which is specific for the two male sterile, 350 and 447, cytoplasms of Vicia faba L. Rearrangement in this male sterile specific plasmid-like DNA molecule seems to produce the 1540 bp plasmid-like DNA molecule found in one line bearing the male sterile cytoplasm.

By hybridization study, the two mitochondrial plasmid-like DNA molecules of 1700 bp show strong homology between themselves. Only the plasmid-like DNA molecule of 1700 bp found in all the cytoplasms shows a weak homology with the plasmid-like DNA molecule of 1420 bp.

The small supercoiled DNA molecules didn't share any homology neither with the chloroplastic DNA of faba bean, neither with the mitochondrial DNA of faba bean, maize and sunflower. But they share homology with the nuclear DNA of maize and with one sequence found in some but not all nuclear genomes of faba bean.

REFERENCES

1. Kemble, R.J. and Bedbrook, J.R. (1980) Nature 284, 565-566.
2. Goblet, J.P., Boutry, M., Duc, G. and Briquet, M. (1983) Plant Mol. Biol. 2, 305-309.
3. Negruk, V.I., Cherny, D.I., Nikiforova, I.D., Aleksandrov, A.A. and Butenko, R.G. (1982) FEBS Lett. 142, 115-117.
4. Wahleithner, J.A., Bailey, J.N., Leroy, P. and Wolstenholme, D.R. (1983) J. Cell Biol. 97, 368a.
5. Powling, A. (1981) Mol. Gen. Genet. 183, 82-84.
6. Hansen, B.M. and Marcker, K.A. (1984) Nucl. Acids Res. 12, 4747-4756.

CLONING AND ANALYSIS OF THE AUTONOMOUSLY REPLICATING SEQUENCES *a* AND *ars*B OF PETUNIA HYBRIDA CHLOROPLAST DNA.

J.M. de HAAS, N. OVERBEEKE, H.J.J. NIJKAMP AND A.J. KOOL

1. INTRODUCTION

Vectors for plant cell organelles may provide a means for tra fer of specific cytoplasmic properties. As a first step towards construction of such vectors, we are attempting to identify and isolate the cpDNA replication origin(s) by cloning cpDNA sequenc that are able to replicate autonomously in yeast.

2. RESULTS

Based upon the assumption that the origin of replication of chloroplast DNA will function as an autonomously replicating se- quence (ARS) we have isolated cpDNA sequences that promote repli tion in yeast by cloning in the specialized yeast vector YIp5 (REF. 1). All transformed yeast cells contained plasmids with cpDNA inserts (REF. 2).

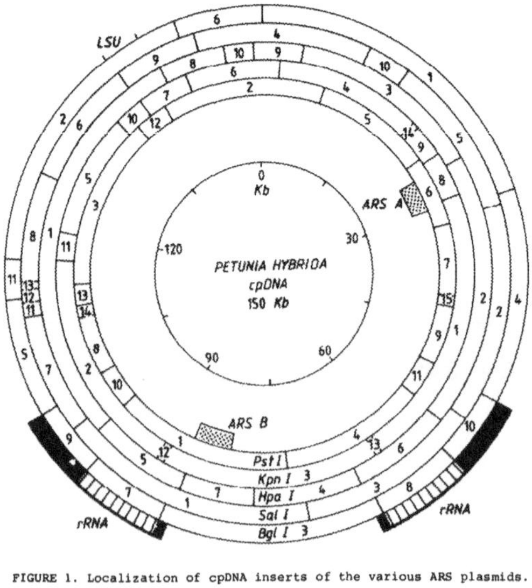

FIGURE 1. Localization of cpDNA inserts of the various ARS plasmids.

Table 1. Replication of *P. hybrida* chloroplast ARS plasmids in *E. coli* and *Anacystis nidulans*.

Plasmid	Replication origin	*E. coli* wt	*E. coli* *polA*$^-$	*E. coli* *polA*$^+$/*ura*$^-$	*A. nidulans*
pLAFR	Pol A indep.	+	+	+	-
pUC303	*A. nidulans* + pBR322	+	-	+	+
pPCY20	*ars* A + pBR322	+	-	+	-
pPCY20-9	*ars* A	+	-	-	-
pPCY51	*ars* B	+	N.D.	N.D.	N.D.

A number of plasmids had an Eco RI insert of 5.4 Kb, which was localized on a position between 25 and 31 Kb on the chloroplast genome (Fig. 1). This region is denoted *ars*A. The inserts of the other plasmids hybridized with a 5.8 Kb Hind III cpDNA fragment that is localized in the small single copy just outside the left inverted repeat. This region is denoted *ars*B.

To determine whether these ARS sequences can function as an origin of replication in prokaryotic organisms, the ARS plasmids were transformed to *polA*$^-$ *E. coli* cells and to the blue algae *Anacystis nidulans*. After removing the pBR322 origin of replication, the remaining *ars*A plasmid pPCY20-9 was also transferred to *polA*$^+$/*ura*$^-$ *E. coli* cells (Table 1). The results indicate that the *ars*A plasmids could neither transform the *A. nidulans* cells nor the *E. coli* *pol A*$^-$ or *pol A*$^+$ cells.

The most intriguing question at this point is whether an ARS represents an origin of replication in the chloroplast *in vivo*. We are using an organelle-free DNA synthesizing system, prepared from isolated chloroplasts to answer this question (REF. 3). The most direct evidence, however will be the reintroduction of the ARS into the chloroplast (e.g. by microinjection or transformation), followed by the demonstration that the ARS sequence is stable maintained as a result of autonomous replication.

REFERENCES

1. D.T. Stinchcomb, M. Thomas, J. Kelly, E. Selker and R.W. Davis. 1980. Proc. Natl. Acad. Sci. 77: 4559.
2. N. Overbeeke, M.A. Haring, H.J.J. Nijkamp and A.J. Kool. 1984. Plant Molec. Biol., 3: 235.
3. N. Overbeeke, J.H. de Waard and A.J. Kool. 1984. In: Proteins involved in DNA replication (U. Hübscher and S. Spadari, eds.). Plenum Press, London, in press.

HIGH MOLECULAR WEIGHT POLYPEPTIDES OF IN-ORGANELLO MITOCHONDRIAL PROTEIN SYNTHESIS IN PETUNIA HYBRIDA.

F. VAN DER MARK AND A.J. KOOL

1. INTRODUCTION

Cytoplasmic male-sterility (CMS) is of importance for commercial hybrid seed production because it prevents self-fertilization of the seed parent plant. At this moment CMS is the only example of a variant phenotype resulting from variations in the plant mitochondrial genome (REF. 1). Intact mitochondria isolated from plant material are able to carry out protein synthesis, the products can be analyzed by gelelectrophoresis. This method is used to study the presence of variant polypeptides associated with CMS phenotypes. So far mitochondria for in-organello protein synthesis have been isolated from etiolated shoots or in some cases from green leaves. Here we describe the isolation of mitochondria from suspension cultured cells to study mitochondrial protein synthesis in Petunia.

2. METHODS

Mitochondria were isolated from Petunia hybrida cell-suspension cultures (300 ml) with a packed cell volume (pcv) of 60 ml. The cells were disrupted with an ultrathurrax or a Bead beater. Purification of the mitochondria was performed by differential centrifugation and discontinuous sucrose gradients or self-generating gradients of Percoll in combination with linear gradients of polyvinylpyrrolidon-25 (REF. 2).

3. RESULTS AND DISCUSSION

It was found that the efficiency of homogenization of plant cells from cell cultures is higher when instead of the ultra thurrax the Bead beater is used: the amount of isolated mitochondria is enhanced about 4-fold. Moreover, by using the Bead beater, the percentage of intact mitochondria as estimated by the cytochrome C oxidase assay increased from 50 to 70%. By using Percoll/PVP-25 gradients (REF. 2) isolated mitochondria with an improved rate of protein synthesis were obtained.

Analysis of proteins synthesized *in vitro* by mitochondria from several plant species show essentially similar patterns of products ranging from 8 to 54 kd. We observed that mitochondria isolated from several Petunia cell lines synthesize polypeptides of high molecular weight (>54kd). To exclude a possible involvement of plastids, control experiments performed by using several energy systems (REF. 3) and purification procedures (REF. 2) were negative. The results of these experiments indicate that the high molecular weight polypeptides are indeed synthesized by the mitochondria.

Although we have not yet examined the pattern of polypeptides synthesized in Petunia leaf mitochondria our results obtained so far suggest that differential expression of the mitochondrial genome occurs. Indications for tissue- and developmental stage-specific expression of the mitochondrial genome have also been obtained in other plant species such as in albino tissue of the iojap mutant (REF. 3). Also specific differences between polypeptides synthesized by mitochondria from etiolated or green tissue were shown (REF. 4).

REFERENCES

1. C.J. Leaver and M.W. Gray. 1982. Annu.Rev. Plant Physiol. *33*, 373-402.
2. D.A. Day, M. Neuburger and R. Douce. 1984. In press.
3. R.J. Newton, G. Zabala, A. Maloney, T. Oro and V. Walbot. 1984. Abstract EMBO workshop p. 30, Melrose.
4. M. Boutry, A. Faber, M. Charbonnier and M. Briquet. 1984. Plant Molec. Biol., In press.

A PERUVIAN PROGENY AS A SOURCE OF A NEW MITOCHONDRIAL VARIABILITY IN MAIZE

A. MEZZAROBBA, J.C. NICOLAS, A BERVILLE and A. CORNU
Station d'Amélioration des Plantes, I.N.R.A., DIJON France

From a population of Peruvian maize, cultivated in greenhouse conditions, one plant was pollinated by the french early line F7. In the offspring of this hybrid, a plant (P2) was recognized as susceptible to methomyl by leaf test bioassay. By backcrossing this material by F7 or F215 inbred (as male recurrent parent), some progenies were obtained that were remarkable by several aspects :

1°) After three back-crosses by F7, most of the progenies appear male-sterile and susceptible to *Helminthosporium maydis race T (HmT)* toxin and methomyl. These traits are characteristic of Texas cytoplasmic male-sterility. However the male fertile plants obtained among the segregants did not restore the male sterility of F7-T line.

2°) Part of this material, crossed at least 2 times by F215 (CMS-T maintainer), gave different results : most of the plants were male-fertile and segregated for resistance/susceptibility to *HmT* and methomyl.

3°) A mt-DNA analysis of P2, P4 and P5 hybrid progenies (Peruvian x F7), and several offspring from P2, has been started, according to the methodology previously described (1) (2). Bgl I, Sal I and Xho I restriction enzymes were used as shown in the figure 1. Obviously, P4 exhibits a pattern, whatever the enzyme used, never described for maize mt-DNA : the two large bands at 9 and 5,5 Kb are typical. P5 looks like T after digestion by Sal I (not shown) but appears different after digestion by Xho I (see figure) ; furthermore P5 is resistant both to methomyl and *HmT* toxin. In contrast P2, which is susceptible to both compounds exhibit a pattern that is quite different with the three restriction enzymes. Moreover, P2 progenies Pa and Pd, for example, are different from P2 origin (with Xho I and Sal I), which in several àspects look like N patterns (Bgl and Sal I). However, in other progenies showing the same phenotype, P2 pattern does not appear to be modified.

It is too early to explain why in this population the mt-DNA is so unstable. The cross by the french line may have initiated rearrangements or partial amplifications within the mitochondrial genome. The mt-DNA analysis of our material from back-crossing by F7 or F215 could permit us to establish whether the male parent is involved in such mt-DNA variability. Nevertheless, it seems evident that in this maize population could coexist several patterns of mt-DNA, similar or different from the T, C, N (seen figure) or S (not shown here) types previously described.

REFERENCES

1. VEDEL F., QUETIER F. 1978. Hydrolyse specifique de l'ADN chloroplastique et de l'ADN mitochondrial des végétaux supérieurs par les enzymes de restriction. Physiol. Vég., 16, 411-425.
2. BERVILLE A., PAILLARD M. 1982. Comment améliorer chez le maïs (*Zea may* L.) la résistance du cytoplasme Texas à *Helminthosporium maydis* race T et *Phyllosticta maydis*. Les colloques de l'INRA, 11, 125-136.

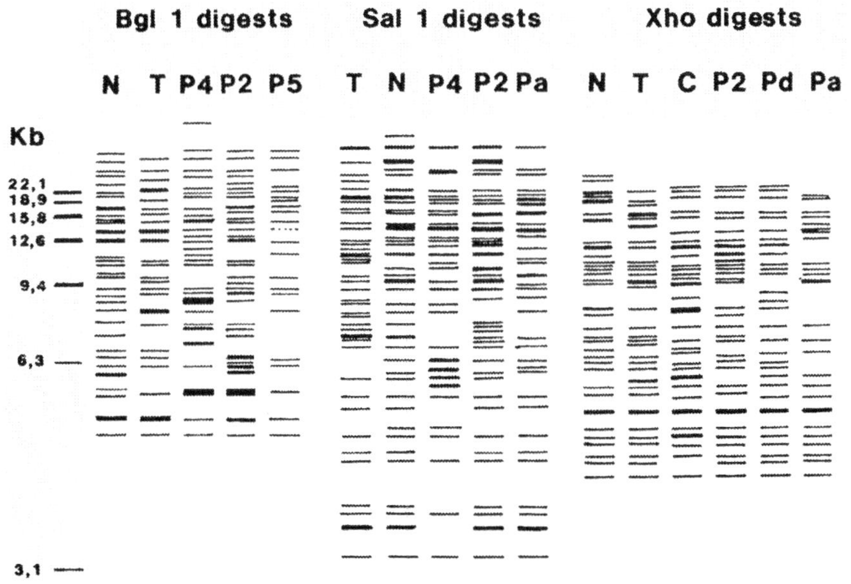

FIGURE 1. Diagrams of Bgl 1, Sal 1 and Xho 1 digests of maize mitochondrial DNA-N, T, C, respectively Normal, Texas and Charrua cytoplasms. P2, P4 and P5 are sister progenies of the Peruvian plant. Pa and pd are progenies of the P2 plant.

THE EFFECT OF HEAT AND CHEMICAL TREATMENTS ON MERISTEM AND SHOOT TIP CULTURE OF CMS AND FERTILE SUGAR BEET.

D.R. DAVIES AND A. POWLING
JOHN INNES INSTITUTE, COLNEY LANE, NORWICH NR4 7UH, ENGLAND

The mt DNA of cms forms of sugar beet differs from that of fertile genotypes in the nuclease digest pattern of the main genome, and in the number and molecular weight of the supercoiled DNA circles which are associated with the main genome. There are no obvious differences in the nuclease digest patterns of the ct DNA of the two fertility classes. (see Powling and Ellis, 1983). This suggests that the genetic control of cms resides in the mt DNA, although there is no proof of this.

We have tried to interconvert the two forms by appropriate treatments of cultured shoots and meristems. Lichter (1978) reported the conversion of cms sugar beet to male fertile by treating meristems at high temperature, and Curtis (1967) showed that interconversions were also possible; by grafting a fertile scion on to a cms stock, a proportion of the fertile scions became cms.

Shoot cultures of a cms and a male fertile genotype were multiplied in vitro to provide uniform genetic material. These were then exposed to chemicals likely to mutate organelle DNA; streptomycin at 10ppm for 6 weeks, ethidium bromide at 10ppm for 1 week in the dark, mitomycin at 12ppm for 2 weeks in the dark. All treatments were at 25°C.

The concentrations of the chemicals used and the length of the treatments were chosen to be just below the levels which gave a significant mortality of the shoots in culture. In addition a small number of shoots were treated with acridine orange (10, 30, or 100ppm) or tetracycline and tylosin (both at 10, 30, or 100ppm) for 4 weeks at 15°C, 16hr daylength. After treatment the shoots were transferred to a standard medium to recover, and rooted prior to transplanting to soil in the greenhouse. They were later planted in the field and scored for fertility. A total of 981 plants were treated together with 317 controls. No interconversion of cms and male-fertile was observed.

Another series of treatments involved heat treatment of meristems followed by shoot tip isolation and culture. Shoots were incubated at 37°C (16hr) day and 28°C (8hr) night. The shoots of the cms form survived the 9 weeks of treatment but none of the fertile line survived beyond $6^1/2$ weeks. At weekly intervals during the treatment some shoots were removed and the apical tips (approx. 0.5 mm) were dissected and incubated at 20°C (16hr daylength). The shoots derived from these tips were rooted and planted as previously described. 123 cms shoot tips survived but only 40 male fertiles. No change in fertility class was observed.

REFERENCES

Curtis, GJ. 1967. Graft-transmission of male sterility in sugar beet (Beta vulgaris L.). Euphytica 16, 419-424.
Lichter, R. 1978. The restoration of male fertility in cytoplasmic male-sterile sugar beet by heat tretment and meristem culture. Z. Pflanzenzucht. 81, 159-165.
Powling, A. and Ellis, THN. (1983). Studies on the organelle genomes of sugarbeet with male-fertile and male-sterile cytoplasms. Theor. Appl. Genet. 65, 323-328.

Microanalysis of variant mitochondrial polypeptides associated with the cytoplasmic male sterility trait in faba bean.

M. BRIQUET, A.M. FABER, P. BERTHELEM[1], J. LE GUEN[1], and G. DUC[2].
Laboratoire d'Etude de l'Hérédité Cytoplasmique, Université de Louvain, Place Croix du Sud, 1, B-1348 Louvain-la-Neuve, (Belgique)
[1] Station d'Amélioration des Plantes, INRA, 35650 Le Rheu (France)
[2] Station d'Amélioration des Plantes, INRA, 21034 Dijon (France)

1. Introduction

The maternally inherited trait termed cytoplasmic male sterility is the main exception of a non-lethal mitochondrial mutation presently known in higher plants. This mutation responsible for the abortion of pollen by an unknown biochemical mechanism, is a powerful tool for the commercial production of hybrid seeds. There are now several biochemical data that support the mitochondrial inheritance of this mutation. Among them the presence of mitochondrial polypeptides the synthesis of which are under the control of nuclear restorer genes that suppress the male sterile phenotype has been described in maize and is of major importance for the study on the biochemical mechanism of pollen abortion (1).

In order to bring additional experimental data to attempt to answer, in this field, to the question of the relation between the presence of variant polypeptides and the male sterile phenotype and to the more general question of nucleo-cytoplasmic interactions we have analysed by a rapid and highly resolutive microprocedure, the mitochondrial translation products of fertiles, male-steriles, restored and revertant lines of faba beans.

2. Material and Methods

Seeds of faba bean (*Vicia faba* L.) have been provided by Dr. P. Berthelem and J. Le Guen, Station d'Amélioration des Plantes de l'INRA, Le Rheu (France) and Dr. G. Duc, Station d'Amélioration des Plantes de l'INRA, Dijon (France).

The micromethod developped for the *in vitro* labelling of mitochondrial polypeptides by ^{35}S-methionine followed by SDS-PAGE-gel electrophoresis and autoradiography has been described previously (2).

3. Results

The mitochondrial translation products of two lines of faba bean in which the same nucleus (G58-B) has been transferred in the two male sterile cytoplasms 447 and 350, have been analysed.

The results show that the synthesis of the 24 000 Mr variant polypeptide is reduced in the 447 cytoplasm if compared with the synthesis of the variant polypetptides when the nucleus G58-B is transferred in the 350 cytoplasm. In parallel it must be noted that the sterility phenotype of this G58-B/447 line is poor in comparison to the G58-B/350 line. Therefore it appears that there is a correlation between the extent of sterility and the extent of synthesis of the variant polypeptide. Our results also show that a polypeptide of 26 000 Mr present in the fertile lines, is also synthesized in the male sterile cytoplasm but at a lower concentration.

We have also analysed the mitochondrial translation products of restored lines that contain restorer alleles derived from the HG-115 line in the context of the two male sterile cytoplasm 447 and 350. In the presence of the nuclear background G58-B the synthesis of the 24 000 Mr polypeptide is more strongly suppressed than in the presence of the nucleus Ad23. In parallel good restoration of the synthesis of the 26 000 Mr polypeptide is observed in the presence of the G58 nuclear background.

These results are well correlated with the degree of restoration of the sterility phenotype. Revertant lines of the two male sterile cytoplasms exhibit also a decreased synthesis of the 24 000 Mr male sterile variant polypeptide and the reappearance of the 26 000 Mr polypeptide.

4. Conclusion

The results presented here support the involvement of the variant mitochondrial polypeptide of 24 000 Mr in the male sterility phenotype of faba bean and illustrate by biochemical observations the nuclear-mitochondrial interactions in higher plants.

5. References

1. Forde, B.G. and Leaver, C.J. (1980) Proc. Natl. Acad. Sci. U.S.A. 77, 418-422.
2. Boutry, M., Faber, A.M., Charbonnier, M., and Briquet, M. (1984) Plant Mol. Biol. (in press).

In vitro study of mitochondrial protein synthesis in cytoplasmic hybrids of *Nicotiana tabacum*.

M. CHARBONNIER, A.M. FABER, G. PELLETIER[1], G. BELLIARD[2], M. BOUTRY and M. BRIQUET
Laboratoire d'Etude de l'Hérédité Cytoplasmique, Université Catholique de Louvain, Place Croix du Sud, 1, B-1348 Louvain-la-Neuve, Belgique.
[1] Laboratoire de Biologie Cellulaire, CNRA, F-78000 Versailles, France;
[2] Laboratoire de Génétique Végétale, Institut de Botanique, 28 rue Goethe, F-67083, Strasbourg Cedex, France.

1. Introduction

Within the framework of the study of the organization and the expression of the extranuclear genetical information in plants, we study the mutation of cytoplasmic male sterility which is located in the mitochondrial DNA. We study the products of the *in vitro* mitochondrial protein synthesis of regenerated tobacco plants from fused protoplasts which have been produced by Belliard *et al.* (1977).

They used 2 varieties of tobacco differing by nuclear genes controlling the leaf shape and by cytoplasmic factors responsible for the flower shape and male sterility or fertility. Cytoplasmic hybrids have been obtained from fusion between protoplasts of *Nicotiana tabacum* var. *Xanthi*, which has normal fertile flowers and petiolated leaves and *Nicotiana tabacum* var. *Techné* which has abnormal flowers with cytoplasmic male sterility, producing no pollen and sessile leaves. Cytoplasmic hybrids (= cybrids) of various phenotypes, distinguishable from the somatic nuclear hybrids and parental types have been identified among the regenerated plants. Genetic analysis has showed that the particular characters of these cybrids are maternally inherited and that they are stable through the sexual cycles. According to the modification of the flower morphology (shape of corolla, stamen filament and anther) the cybrids have been sorted out in 4 classes.

The male fertile plants possess a nuclear *tabacum* genome associated with a *tabacum* cytoplasm. The male sterile plants are characterized by a nuclear *tabacum* genome inside a *debneyi* cytoplasm. The cybrids have a nuclear genome of one of the two parents associated with an hybrid cytoplasm *tabacum* x *debneyi*.

Analysis with restriction endonuclease has shown that only one or the other parental chloroplast DNA is present in the progeny of the cybrids (Belliard *et al.*, 1978).

Analysis with the restriction endonuclease SalI has shown that the mitochondrial DNA of cybrids are different from those of the parents and from the mixture of the two. The new DNAs result from mitochondrial recombination (Belliard *et al.*, 1979).

2. Methods

Mitochondria are isolated from young green leaves as previously described by Boutry *et al.* (1984) with light modifications. The *in vitro* mitochondrial protein synthesis has been performed with ^{35}[S]-methionine. The products of the mitochondrial protein synthesis are analysed by SDS-PAGE electrophoresis and autoradiographied.

3. Results

The autoradiography of the synthesized polypeptides in the mitochondria from the male sterile line is characterized by the appearance of a polypeptide of M.W. 18.000 and by disappearance of a polypeptide of M.W. 17.000, which is present in the male fertile line. Among the 8 studied cybrids, the synthesized polypeptides in 2 of them (T13 and X7) are identical to the male sterile line and the synthesized polypeptides in the 6 others are identical to the male fertile line.

T13 and X7 are cybrids from the class IV, which are the most sterile whereas the others are cybrids from the class I to III. It is the first time that a correlation between the cytoplasmic male sterility and the presence of a variant mitochondrial polypeptide is clearly established. There is no correlation between the SalI band patterns of the mitochondrial DNA and the presence or the absence of these polypeptides.

4. References:

- Belliard, G., Pelletier, G., Férault, M. (1977) C.R. Acad. Sci. 284, 749-752.
- Belliard, G., Pelletier, G., Vedel, F., Quetier, F. (1978) M.G.G. 165, 231-237.
- Belliard, G., Vedel, F., Pelletier, G. (1979) Nature 281, 401-403.
- Boutry, M., Faber, A.M., Charbonnier, M., Briquet, M. (1984) Plant Mol. Biol. (in press).

HYBRID CELL LINES DAUCUS CAROTA x ORYZA SATIVA: EVIDENCE FOR THE SELECTION OF "ASYMETRIC" HYBRID GENOMES.

Sala F., Biasini M.G., Morandi C., Nielsen E., Parisi B. and Sala C.
Dipart. di Genetica e Microbiol., Univ. di Pavia, Italy.

In order to study the fate of the parental genomes in somatic cell hybrids between distantly related species, protoplasts from cultured cells of Daucus carota and Oryza sativa have been fused according to Schieder and Vasil (1). Selective conditions were based on the inhability of rice protoplasts to grow in a culture medium optimal for carrot protoplasts and on the naturally occurring resistance displayed by rice protoplasts to the proline analogue azetidine-2-carboxylic acid (A2CA) (2). Selection resulted in hybrid colonies which combined both traits. Hybrids morphologically resembled carrot cells and were characterized by generation times longer than those of the parents and by variable chromosome numbers with a mode from 20 to 48.

One of these lines (CR1), showing high resistance to A2CA, was chosen for a molecular analysis of the nuclear genome.

Fractionation on agarose gel of the nuclear DNAs purified from the parental and from the hybrid cell lines, after treatment with the restriction enzyme BamH1, resulted in a distribution of repeated sequences which showed strong homologies between hybrid and carrot cells. Only a few rice specific sequences were observed. Furthermore, a procedure for measuring the relative contribution of the parental cells to the hybrid nuclear genome has been developed.

This is based on molecular hybridization of nuclear DNA from parental and hybrid cell lines bound to nitrocellulose filters, with nick-translated nuclear DNA from rice nuclei. This approach is feasible owing to the low homology between rice and carrot DNA (Fig.1). In the case of the CR1 hybrid, the contribution of the rice genome to the hybrid nuclear DNA was shown to be about 7%, if the assumption was made that in the hybrid line no significant variation occurred in the percentage of carrot genome homologous to rice DNA.

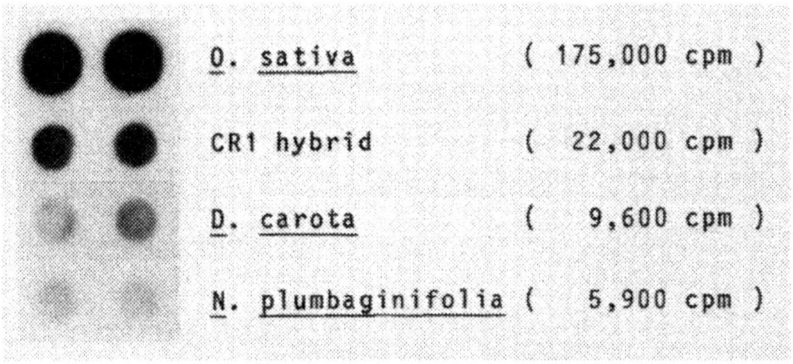

Fig.1- Spot hybridization of different nuclear DNAs with ^{32}P-labelled rice nuclear DNA.

To obtain information about the homologies between the hybrid and parental lines at the translational level, soluble protein was extracted and fractionated by polyacrylamide gel electrophoresis. Also in this case the results (Fig.2) showed strong but not complete homologies between the proteins of the CR1 hybrid and carrot cells, with a few rice specific proteins possibly present in the hybrid cells.

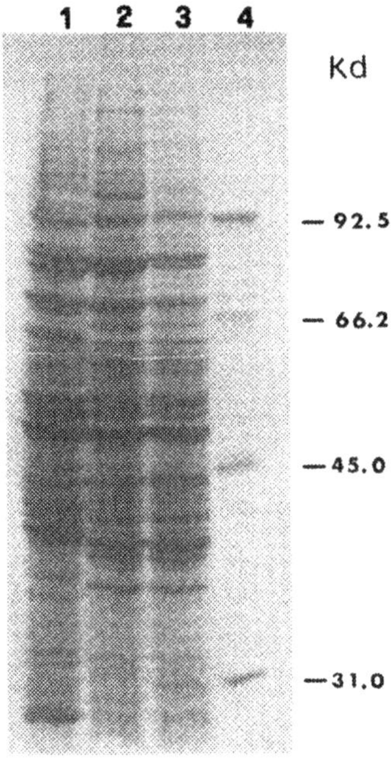

Fig.2- Electrophoresis in a 1.0% polyacrylamide gel of total soluble proteins. 1, rice; 2, CR1 hybrid; 3, carrot; 4, marker proteins.

REFERENCES

1. Schieder O. and Vasil I.K. (1980). Inter. Rev. Cytol., suppl. 11B, 21.
2. Nielsen E., Rollo F., Parisi B., Cella R. and Sala F. (1979). Plant Sci. Lett. 15, 113.

MOLECULAR STUDIES ON SOMATIC HYBRIDS OF DATURA INNOXIA WITH ATROPA BELLADONNA, PHYSALIS MINIMA AND VARIOUS DATURA SPECIES

E. MÜLLER-GENSERT, J. STOCKHAUS and O.SCHIEDER, Max-Planck-Institut für Züchtungsforschung, D-5000 Köln 30, FRG

1. INTRODUCTION

Two different methods have been used to determine the composition of genetical material in somatic hybrids of Datura (1)

a) Restriction fragment length polymorphism (RFLP) (2), to determine chloroplast segregation (3) and detection of nuclear material specific to each parent in the hybrid (4).
b) Use of species-specific repetitive DNA-clones (5) to assay the nuclear constitution of the hybrid.

2. PROCEDURE

Plant DNA was isolated from leaf tissue of greenhouse grown plants or from sterile callus cultures, according to standard methods (6).
Plant DNA was digested with restriction enzymes, blotted to nitrocellulose, and hybridized to radioactive cloned DNA (7). Potato chloroplast DNA was used to probe for chloroplast segregation, whereas the nuclear constitution was determined by probing with nuclear DNA of different genetic origin. The banding pattern after exposure to X-ray film allows conclusions about the genetic composition of the hybrid. The method to obtain species-specific clones has been described by Saul and Potrykus (5). These clones were selected, blotted on nitrocellulose and hybridized to radioactive DNA of the respective hybrids.

3. RESULTS

3.1. Chloroplast segregation

Segregation in the different hybrids examined is displayed in table 1:

1st parent	No.	plastome	1st parent	No.	plastome
A.belladonna	1	D.innox.	D.stramonium	1	D.innox.
P.minima	1	D.innox.		2	D.stram.
D.quercifolia	1	D.innox.		3	D.innox.
D.sanguinea	1	D.innox.		4	D.innox.
D.discolor	1	D.innox.		5	D.innox.
	2	D.innox.		6	D.innox.
D.candida	1	D.innox.		7	D.innox.
	2	D.cand.			

Table 1: Chloroplast segregation in somatic hybrids of *Datura*. In all cases the 2nd parent is *Datura innoxia*.

3.2. *Nuclear constitution assayed by RFLP*.

Preliminary results show the suitability of this method. However, screening for the right clone-enzyme combination is time consuming and often impeded by partial digestion of the plant DNA, which gives rise to irreproducible results.

3.3. *Use of species-specific clones*.

Currently 600 clones from *A. belladonna* and 800 clones from *D. innoxia* have been obtained. Out of the *Atropa* clones 30 were found to be specific, whereas 24 *Datura* clones exhibited specificity. All of the specific clones hybridized to DNA of the respective hybrids. Considering that one of the hybrids contains only two to four chromosomes of *Atropa*, we assume that the specific Atropa clones are either sibling clones from one or more of these chromosomes or that they represent a family of repetitive DNA that is spread all over the genome.

4. REFERENCES

1 Schieder O, Gupta PP, Krumbiegel G, Hein T. 1983. Ciba Foundation Symposium No. 97, London
2 Beckmann JS, Soller M. 1983. Theor. Appl. Genet. 67, 35-43
3 Fluhr R, Aviv D, Galun E, Edelman M. 1983. Theor. Appl. Genet. 67, 491-497
4 Uchimiya H, Ohgawara H, Kato H, Akiyama T, Hamada H. 1983. Theor. Appl. Genet. 64, 117-118
5 Saul MW, Potrykus I. 1984. Plant Cell Rep. 3, 65-67
6 Bedbrook J. 1981. PMB Newsletters IIB 24
7 Southern EM. 1975. J. Mol. Biol. 503-517

STUDIES ON THE TRANSFER OF GENETIC INFORMATION BETWEEN TOBACCO AND POTATO BY PROTOPLAST FUSION.

D. FOULGER, N. FISH, S.W.J. BRIGHT, M.G.K. JONES.
Rothamsted Experimental Station, Harpenden, Herts., U.K.

As part of a study on gene transfer in crop plants by protoplast fusion, work on introduction of defined genes into potato has concentrated on transfer of resistance to S-2-aminoethyl-L-cysteine (Aec) from the *Nicotiana sylvestris* mutant RAEC-1 (Negrutiu *et al*., 1984) to British potato (*Solanum tuberosum*) cultivars .

In order to identify heterokaryons, protoplasts of different origin have been used - colourless suspension protoplasts of *N. sylvestris* and green mesophyll protoplasts of *S. tuberosum* derived from shoot cultures (cvs. Maris Piper, Desiree, King Edward). Suspension and mesophyll protoplasts have been fused using polyethylene glycol (MW 6000) and Ca^{2+}/high pH treatment. Heterokaryons were identified by the presence of chloroplasts in cytoplasm typical of suspension cells. This identification has been confirmed by fluorescence after labelling suspension protoplasts with fluorescein isocyothionate. Heterokaryons fluoresced yellow-green from fluorescein and red from chloroplasts. Of 1500 heterokaryons isolated by micropipette 3 days after fusion, and grown on using nurse cultures of immobilised potato protoplasts, 24 colonies developed and 18 were resistant to 10 mgl^{-1} Aec. Putative hybrid callus lines are being tested for expression of 7 isoenzyme activities. Clear and useful differences in isozyme patterns between the partners have been found for 6-phosphogluconate dehydrogenase, glutamate dehydrogenase and esterase. The three lines analysed in detail for these isozymes have shown only the *N. sylvestris* patterns. Regenerated plants also resembled *N. sylvestris*. We are confident that the lines tested were interspecific heterokaryons when isolated, and it therefore appears that part or all of the potato genome has been eliminated during culture.

Protoplasts of *N. sylvestris* have also been inactivated by X- and γ-irradiation (5-100 krad) (in all experiments except one, these dose levels prevented colony formation) before fusion with potato. Following low irradiation levels (5-10 krad), several hundred Aec-resistant colonies were recovered (see Table 1). Nine colonies analysed so far for one isoenzyme showed only the N. sylvestris bands. These results suggest that the irradiated N. sylvestris nuclei recover within the heterokaryons and again eliminate potato components. We are testing the hypothesis that by increasing irradiation doses to N. sylvestris, the proportion of potato to N. sylvestris genomes can be increased to a state where hybrid regenerants contain essentially the potato genome with limited contributions from N. sylvestris.

Table 1. Aec-resistant colonies recovered following irradiation of N. sylvestris protoplasts .

	X-ray dose (krad)		
Protoplast source, numbers	1.5	5	10
S. tuberosum ($X10^5$)	12.95	9.93	21.7
N. sylvestris ($X10^5$)	4.57	10.4	17.7
Aec-resistant colonies	715	231	40

In addition to these experiments work is also in progress to fuse potato with Nicotiana sp carrying different marker genes, and with other Solanum sp using appropriate selection systems to allow the isolation and culture of hybrid callus.

Negrutiu I, Cattoir-Reynearts A., Verbruggen, I., Jacobs, M. (1984). Lysine overproducer mutants with an altered dihydrodipicolinate synthase from protoplast culture of Nicotiana sylvestris (Spegazzini and Comes). Theor. Appl. Genet 68, 11-20.

FUSION OF CROP PLANT PROTOPLASTS IN HOMOGENEOUS ELECTRIC FIELDS FOR ANALYTICAL AND PREPARATIVE PURPOSES.

M.J. TEMPELAAR and M.G.K. JONES
Rothamsted Experimental Station, Harpenden, Herts., U.K.

The inhomogeneous electric field fusion system described by Zimmermann (1982) utilises a high frequency non-homogeneous alternating field to line up protoplasts in chains, followed by a short direct current pulse to induce protoplast fusion. We have modified this approach to use an homogeneous alternating field with a larger separation between the electrodes (Tempelaar and Jones, 1985).

When applied analytically, membrane breakdown characteristics, which determine fusion and uptake capabilities of different kinds of protoplasts, can readily be investigated by scoring fusion frequencies in the aligned protoplasts under the light microscope. Energy dose-fusion response curves have been obtained that show major differences between mesophyll protoplasts (_Solanum brevidens_, _Triticum aestivum_) and protoplasts from suspension cultures, (_S. brevidens_, _Nicotiana rustica_, _N. silvestris_, _Datura innoxia_, _T. aestivum_), root-tips (_Vicia faba_) and hypocotyls (_Brassica napus_), Table 1, a and b.

This kind of analysis may be used in various ways, for instance to determine rapidly differences in fusogenic responses under different culture conditions, to establish the effect on electrofusion of temperature, ions and compounds used in chemical fusion techniques and to determine conditions for uptake of particles, DNA and plasmid vectors.

On a preparative scale, differences in fusogenic responses of protoplasts in mixtures have been used to control the course of the fusion process. After studying the dose-response curves, electrical fusion parameters have been selected to initiate the fusion process only in the most sensitive partner (e.g. leaf protoplasts). When present at a lower density than the less sensitive protoplasts (e.g. suspension protoplasts), most of the aligned leaf protoplasts fused with suspension protoplasts, Table 1, c. Up to 60-70% of the fusion products originated from 1:1 fusions at 40-50% overall fusion frequencies. Numerous colonies have been obtained following electrofusion and are being cultured.

Without using the origin-dependant differences in characteristics, 1:1 heterologeous fusion products cannot be obtained in a similar selective fashion. Thus for preparative electrofusion there are clear advantage in using combinations with natural (e.g. mesophyll-suspension) or induced differences in fusion characteristics) in terms of heterokaryon yield. In addition it might be envisaged that other properties of different types of protoplasts (e.g. density, surface properties) might facilitate purification of the fused products for culture, LM and EM studies. Alternatively, conditions may be changed to generate more multi-fusion products, which may be interesting for investigation of intergenomic interactions.

Table 1. Fusogenic properties of leaf and suspension protoplasts

Protoplast source	Fusion frequencies induced by pulse of N electrical units (volts x time) N = 4	N = 10	N = 20
a) *S. brevidens*, leaf	20	75	80
b) *N. sylvestris*, suspension	0	15	30
c) a) x b)	30[1]	50[1]	60[1]

[1]. Fusion frequency expressed as percentage of a) fused to b) in aligned protoplasts.

We believe that the larger capacity of the homogeneous electric field electrofusion system (see Tempelaar and Jones 1985) is better suited to directed 1:1 fusions of plant protoplasts, than the inhomogeneous field system previously described (Zimmermann, 1982) and should facilitate the use of protoplast fusion and uptake techniques for combination or introduction of useful characters within crop plant species.

REFERENCES

1. Tempelaar, M.J. Jones, M.G.K. 1985. Fusion characteristics of plant protoplasts in electric fields. Submitted for publication.
2. Zimmermann, U. 1982. Electric field-mediated fusion and related electrical phenomena. Biochem. Biophys. Acta 694, 227-277.

ELECTRO-FUSION OF PLANT PROTOPLASTS

R. SOTAK, J.B. POWER, AND E.C. COCKING

Plant Genetic Manipulation Group, Department of Botany
University of Nottingham, Nottingham, NG7 2RD, U.K.

Electric field induced fusion of plant protoplasts may be a useful tool in the production of somatic hybrid plants. It has been reported (1) to have several advantages over chemical fusion such as higher fusion frequencies, it can be performed at physiological pH, and occurs in a controlled and synchronous manner.

Electro-fusion is a two stage process. Initially protoplasts are aligned in pearl chains between two parallel electrodes due to their interaction with an alternating, non-uniform electric field. Once membrane-to-membrane contact has been achieved a direct current pulse brings about reversible breakdown of the membranes. Pores form in the membrane contact zone leading to the formation of cytoplasmic bridges between apposed protoplasts. The membranes then reseal and the fused protoplasts round up (2).

A comparative study of chemical and electro-fusion is being undertaken in order to determine whether differences exist concerning fusion frequency, viability of heterokaryons, and throughput of somatic hybrids. Two protoplast systems being studied are: 1) Petunia parodii leaf mesophyll fused with an albino cell suspension of Petunia hybrida, and 2) leaf mesophyll of a nitrate reductase deficient and streptomycin resistant mutant of Nicotiana tabacum fused with a Nicotiana rustica cell suspension. Somatic hybrid plants derived from the above species have been recovered in this laboratory previously through chemical fusion.

The electro-fusion medium normally consists of 0.6 M mannitol with 0.5 mM $CaCl_2$. Protoplasts must be washed carefully in order to minimize the amount of cellular debris that can contribute to Joule heating of the medium and subsequent disruption of the pearl chains. Pearl chain formation is routinely achieved at a field strength of 100 to 200 V/cm. at 0.5 to 1.0 MHz. Lower frequencies may cause rotation of protoplasts and inhibit fusion. The AC field is gradually reduced to zero volts after application of the DC pulse in order to increase the rate of rounding up and to decrease the likelihood of the AC field causing any irreparable damage to the heterokaryons.

Great variation has been observed in the voltages used to induce fusion. Barley, oat, and garlic mesophyll protoplasts fuse readily at high frequencies with pulses of 15-20 volts for 20 usec using platinum or gold electrodes spaced 200 um apart. However when attempting to fuse dicot protoplasts from Nicotiana, Petunia, Linum, Lycopersicon, and Glycine max much higher voltages, 40-50 volts for 50 usec, are needed to bring about even low fusion frequencies, usually less than 1%. When fusion does occur the rounding up time may take 30 minutes or more as compared to as little as one minute for oats and barley. At these higher voltages there is also an increase in the frequency of protoplast lysis.

Electro-fusion of a monocot, such as barley mesophyll, and a dicot, Petunia hybrida cell suspension, occurs at intermediate frequencies of 5-10%. Freeze etching studies of some monocot and dicot species are being undertaken to determine if any structural differences occur in these membranes which may account for the variation in fusion frequency which has been observed.

REFERENCES

1. Zimmermann,U., Vienken, J.(1982) Electric field-induced cell-to-cell fusion. J. Membr. Biol. 67:165-182.
2. Zimmermann, U., Scheurich, P.(1981) High frequency fusion of plant protoplasts by electric fields. Planta 151:26-32.

SOMATIC HYBRIDIZATION IN FORAGE LEGUMES.

D.M. GILMOUR, M.R. DAVEY AND E.C. COCKING.
PLANT GENETIC MANIPULATION GROUP, DEPARTMENT OF BOTANY,
UNIVERSITY OF NOTTINGHAM, UNIVERSITY PARK, NOTTINGHAM.NG7.2RD

Somatic hybridization has the potential to increase variability in forage legumes by the transfer of traits between genera and species which are normally sexually incompatible. The agronomic value of Medicago sativa (alfalfa, lucerne) may thus be improved by somatic hybridization with other Medicago species, or other forage genera. Isozyme analyses of certain accessions of the wild species Medicago falcata have shown alleles not yet observed in any cultivar (1). This indicates the availability of germplasm which may not have been used in alfalfa breeding to date.

Protoplasts have been isolated enzymatically from Medicago species using cell suspensions, leaflets of pot-grown plants, and seedling cotyledons as experimental material. Fusion between M. sativa cotyledon protoplasts and M. falcata cell suspension protoplasts has been carried out using polyethylene glycol (P.E.G.) followed by high pH/Ca^{2+}. Heterokaryons have been identified by labelling cell suspension protoplasts with fluorescein-isothiocyanate (F.I.T.C.) and using the auto-fluorescence of the green cotyledon protoplasts. Fusion products have been selected from homokaryons and unfused parental protoplasts using a micro-manipulator.

Subsequent culture of isolated heterokaryons has been hampered by the fact that cells are only able to grow at certain population densities, which, for higher plant cells, is in the range of 10^4 - 10^6 per ml. It is assumed that this is the result of leaching of cellular metabolites into the medium.

The micro-isolation of sufficient heterokaryons to

to reach this density is not feasible by a manual method, therefore a technique is being developed to increase the plating efficiency of protoplasts at low density.

A simple feeder layer technique has been described (2), for the plating of plant cells and protoplasts at low density. Using cells and protoplasts embedded in agarose, M. sativa cell suspension cells were found to be most efficient at supporting growth of low numbers of M. sativa and M. falcata protoplasts plated in droplets on a cellophane membrane. M. sativa cotyledon protoplasts gave rise to small colonies at a density of 5×10^2 per ml, while M. falcata cell suspension protoplasts would only divide at a density of 5×10^3 per ml. It is hoped that further development of this approach will enable the rescue of cell colonies from individually isolated heterokaryons.

Hybrid callus will be screened biochemically for the expression of M. falcata and M. sativa genes. In particular, the isozyme banding patterns of peroxidases and leucine-aminopeptidase (1), will be used to assess whether M. falcata traits are present in alfalfa.

REFERENCES

1. Quiros, C.F. 1983. Alfalfa, Luzerne (Medicago sativa L.) 253-294. In : Isozymes In Plant Genetics and Breeding, Part B, Edited by S.D. Tanksley and T.J. Orton.
2. Shneyour, Y., Zelcher, A., Izhar, S., and Beckmann, J.S. 1984. Plant. Sci. Lett. 33, 293-302.

PROTOPLAST CULTURE OF FORAGE LEGUMES

K.J. WEBB, M.F. FAY, S. WOODCOCK, L.S. PIKE, P.J. DALE

1. INTRODUCTION

Before we can exchange genetic material between sexually incompatible species of forage legumes, we need: (a) to identify suitable species for use as donors and (b) to control plant regeneration from protoplast-derived callus.

2. IDENTIFICATION OF SUITABLE SPECIES FOR USE AS DONORS

Potential donors of genes for the production of the anti-bloat tannins have already been recognised in *Lotus corniculatus* and *Onobrychis viciifolia*. However, a survey of agronomic characters in a large number of *Trifolium* spp. has revealed several potentially interesting species. For example, in addition to the five clover species known to be bloat-safe, six new species have been identified as containing tannins. Other characters examined were the ability to fix atmospheric N, resistance to the clover rot fungus *Sclerotinia trifoliorum* and response in tissue culture.

The ability to regenerate plants varies according to genotype and species. This characteristic has been studied in a diverse range of forage legumes. Seedling material was used to determine what proportion of the population of each species will regenerate and to select particularly responsive genotypes.

Species vary in their response in culture. In *Trifolium repens*, adventitious shoots were regenerated from approximately 10% of the cultures while the figure was about 8 times higher in *L.corniculatus*. The genotypes which have been selected for their regenerative response are being stored in culture at 4°C for future protoplast isolations and some have been used for seed production.

3. INVESTIGATION OF PLANT REGENERATION FROM COMPLEX EXPLANTS AND PROTOPLASTS

In most species, cotyledons and hypocotyls or hypocotyls-plus-shoots excised from 7 day old seedlings can regenerate *de novo* buds in culture. In some particularly responsive species, such as *L.corniculatus*, cultured roots also form new shoots. Young, fully expanded leaves of the latter species and also *O.viciifolia* regenerate shoots readily but in *T.repens* and *T.hybridum*, only limited shoot regeneration has occurred from this source. The production of roots on these regenerated shoots varies; in general shoots from juvenile material form roots more readily than those from mature tissues. But in *L.corniculatus*, 100% of the shoots regenerated from leaves were rooted and established in soil.

Protoplasts have been isolated and colonies formed from various forage legumes, including from leaf mesophyll cells of *L.corniculatus*. Protoplasts isolated from leaves of both *T.repens* and *T.hybridum* and roots of *T.hybridum* have divided to form callus and a low percentage of these colonies have regenerated shoots. These shoots have formed roots and the plantlets derived from both leaf and root protoplasts of *T.hybridum* have been successfully transferred to soil.

The use of selected genotypes, either as *in vitro* shoots or as soil-grown plants, should increase the frequency of plant regeneration from protoplasts of the less responsive clovers.

THE DEVELOPMENT OF PROTOPLAST SYSTEMS FOR GENETIC EXCHANGE IN FORAGE GRASSES

P.J. DALE, S.J. DALTON, L.A. BOYD, A.D. HALE, J.A. JACKSON

1. INTRODUCTION

For protoplast systems to be evaluated for the genetic modification of crop plants, it is necessary to be able to regenerate plants from protoplasts reliably and repeatably. Towards this goal, various factors are being considered: plant genotype, source of protoplasts, response of protoplasts in culture and the extent of any genetic instability induced by passage through a tissue culture phase.

2. GENOTYPE SELECTION

Plant genotypes have been selected in several forage grass species which readily regenerate plants from callus cultures; frequently by somatic embryogenesis. Further advance is considered possible by this approach, so genotype screening is being continued.

3. SOURCE OF PROTOPLASTS

3.1. Suspension cultures

Protoplasts from mature leaves in the Gramineae are unresponsive, so an alternative source has to be identified. Suspension cultures are attractive for protoplast isolation because they are capable of high growth rates and consist of small cell clumps which aids enzyme penetration. Most attempts at forming suspension cultures in the Gramineae have used callus cultures after a prolonged period of growth on an agar medium which can lead to the loss of morphogenetic potential. Our experiments have concentrated therefore on the establishment of suspension cultures directly from excised embryos. Suspension cultures form rapidly from immature embryos (c. 14 days) but in the perennial grasses, immature embryo production is seasonal. Chopped mature embryos, excised from ripe caryopses, were tested as a source of suspension cultures in the grass species: Lolium perenne,

L.multiflorum, L.temulentum, Festuca pratensis, F.arundinacea, F.ovina, Phleum pratense and Dactylis glomerata. The species L.multiflorum and F.arundinacea reaily formed fast growing fine suspensions (c. 28 days).

The suspension cultures are comprised of polymorphic cell clumps. Experiments are in progress to test their relative potential for plant regeneration and the release and response of protoplasts.

3.2. Explants

Various explants have been tested for their ability to give protoplasts directly. Mature embryos, immature embryos, vegetative meristems and immature leaves were treated with a range of enzyme combinations. Generally the more mature and complex the tissue/ organ the more complicated the enzyme mixture required to release protoplasts. The yield of protoplasts has so far been poor. Immature embryos were studied most and tended to give small delicate protoplasts which burst easily.

4. PROTOPLAST CULTURE

The formatior of callus colonies from rapidly growing suspension cultures of L.multiflorum has been reported previously (1). Callus colonies have now been induced from protoplast cultures of F.arundinacea derived from suspension cultures initiated from mature embryos as described above.

5. GENETIC STABILITY IN CULTURE

For controlled genetic modification it is desirable to be able to understand and manipulate the factors determining genetic and cytological changes in Gramineous tissue cultures. In the initial experiments on diploid L.multiflorum, there was considerable cytological variation. In recent work on the same species, minimal variation was found. The reason for this discrepancy is not yet clear.

REFERENCES

1. Jones MGK, Dale PJ 1982. Z. Pflanzenphysiol. **105**, 267-274.

PROPERTIES OF A RAPIDLY GROWING SUSPENSION CULTURE OF RYE (SECALE CEREALE)

C.M. COLIJN-HOOYMANS, J. SYBENGA

The final aim of this project is the transfer of single and multiple genes to rye and wheat by microinjection of protoplasts or cells. Because microinjection is only an efficient transformation method when applied to small cell clusters or single cells, a culture system for a small amount of cells has to be developed.

To obtain such a system callus cultures were initiated from immature embryos of rye. When 0.5-1 mm embryos were placed on MS medium supplemented with 2 mg l^{-1} 2,4-D vigorous callus growth was observed within 7 days and after 3 weeks a good proliferating callus was obtained. Both loose friable callus tissue and a more compact tissue was produced. This more compact tissue which is white to pale yellow in colour is used to start a suspension culture.

After two months of subculturing on Schenk and Hildebrandt medium (SH) supplemented with 2 mg l^{-1} 2,4-D a rapidly growing cell culture could be obtained consisting of small cell clumps (4-30 cells). The cells can now be transfered every 12 days to fresh medium by a one to five dilution. After 14 days the culture is in the stationary phase.

Plating 2 ml cell culture (10^6 cells/ml) on 24 ml solid SH medium results in callus formation within three weeks. Because it tooks two months before this suspension culture was obtained and because we selected for the most rapidly growing cells the callus obtained from this suspension culture is only capable to induce roots when transfered hormone-free medium. The root differentiation capacity remains present even after one and a half year in culture. From the callus cultures developed from these plated suspension cultures within three weeks similar new suspension cultures could be initiated.

Cell clumps smaller than 10 cells were able to grow in microdroplets and also when they are plated on a feeder layer of this suspension culture. This means that when small cell clumps will be injected they are able to proliferate in low densities. Microinjection of small rapidly dividing cell clumps could be an attractive alternative for the injection of protoplasts. We can isolate protoplasts from these suspension cultures (up to 40% protoplasts of the total amount of cells). By staining with Calco Fluor White cell wall regeneration was observed with 35% of the protoplasts. However, only occasionally a cell division occurred and therefore protoplasts look not a good system for microinjection at the moment.

Further experiments will be focussed on microinjection of small cell clumps and on induction of cell division of these injected cells.

REFERENCES

Schenk, R.U., Hildebrandt, A.C. 1972. Can. J. Bot. 50, 199-204. Medium and techniques for induction and growth of monocotyledonous and dicotyledonous plant cell cultures.

CALLUS FORMATION FROM SUNFLOWER (**Helianthus annuus** L.) PROTOPLASTS

LENEE, P. and CHUPEAU, Y.

1. INTRODUCTION

In breeding programs exploring variability in wild **Helianthus** species, use of protoplasts could prove essential. However, very little information is available on **Helianthus** protoplast culture (1).

Our first results indicate that more stringent conditions are required for sunflower protoplast development compared with other Compositae protoplast culture already established for chicory (2) and lettuce (3,4).

2. MATERIALS AND METHODS

2.1. Growth of plant material: seeds (from I.N.R.A., Clermont-Ferrand, France) of two lines of H. **annuus**, TMA (male sterile) and RHA (male fertile), two hybrids of H. **annuus** "Rodeo" and "Bolero" and a wild species H. **argophyllus**, were either cultivated in the greenhouse or in controlled conditions with various light and temperature regimes, or otherwise germinated **in vitro** (light: 30 $\mu E\ s^{-1}\ m^{-2}$, 16h, 25°C).

2.2. Protoplast isolation: small fragments of cotyledons, hypocotyls, leaves and stems were macerated either in a simple saline medium (KCl, $CaCl_2$) or in the complete T_0 medium (2) of varied osmotic pressure and with various concentrations of Macerozyme R-10, Onozuka R-10 and Driselase. After over-night incubation (dark, 22-25°C) filtered protoplasts were washed by 3 centrifugations (70g, 5 min) in a saline medium of appropriate osmolarity.

2.3. Protoplast culture: washed protoplasts were diluted at initial densities ranging from 5×10^3 to 10^5/ml in a M.S. based medium with varying levels of mannitol, NAA, BA and with one of four different nitrogen sources: 15 mM KNO_3 - 5 mM KNO_3+5 mM NH_4NO_3+10 mM KCl

7.5 mM diammonium succinate+15 mM KCl - 7.5 mM glutamine+15 mM KCl.

3. RESULTS

For all plants tested, weak light (65 µE s^{-1} m^{-2}) and temperatures about 10-20°C appeared to facilitate protoplast isolation from leaves of plants grown in the greenhouse or growth chambers. Osmotic potential given by 0.45 M mannitol in the complete medium is optimal for leaf protoplast obtention and culture. For stems and hypocotyls, however, better yields of viable protoplasts were obtained in the apparently isoosmotic saline medium: KCl 330 mM, $CaCl_2$ 14 mM. In this simple medium with the optimal enzyme mixture (Macerozyme R-10 0.02 %, Onozuka R-10 0.1 %, Driselase 0.05 %) **in vitro** germinated hypocotyls repeatedly yielded $5x10^6$ protoplasts/g fresh weight.

No cell divisions were observed in protoplasts from mesophyll, stems or cotyledons cultured in several liquid media in which the nitrogen source as well as the levels of NAA, BA and mannitol were varied. Only protoplasts derived from hypocotyls began to divide after 4 days in practically all media tested. However, all cultures initiated on media containing nitrate were dead by the end of the third week of culture, even after serial dilution by fresh medium.

After ten days, the best plating efficiency (60%) was obtained in T_0 medium (2) with only a reduced form of nitrogen for initial densities of $4x10^4$ protoplasts/ml. With glutamine or diammonium succinate as sole nitrogen source, serial dilution every 10 days in fresh medium with a reduced amount of NAA finally gave calli transferable onto solid medium from 5% of the initially plated protoplasts.

Varying the levels of various growth substances, sugars and nitrogen sources as well as various light and temperature regimes have not yet made it possible to regenerate plants from calli derived from protoplasts.

REFERENCES

(1) Binding H, Nehls R, Kock R, Finger J, Mordhorst G. 1981. Z. Pflanzenphysiol. 101, 119-130.

(2) Crepy L, Chupeau MC, Chupeau Y. 1982. Z. Pflanzenphysiol. 107, 123-131.

(3) Engler DE, Grogan RC. 1982. Plant Sci. Letters. 28, 223-229.

(4) Berry SF, Lu DY, Pental D, Cocking EC. 1982. Z. Pflanzenphysiol., 108, 31-38.

SELECTABLE MARKERS AND THEIR USE IN THE GENETIC MANIPULATION OF PLANT CELLS : I. ISOLATION AND CHARACTERIZATION OF SEVERAL AMINOACID AUXOTROPHS.

I. Negrutiu and M. Jacobs.
Plantengentica, I.M.B., S[t] Genesius Rhode 1640 BELGIUM

We describe an empirically deviced BUdR enrichment procedure using protoplast cultures of haploid N. plumbaginifolia. The major features of the system are as follows :
(1) The protoplast culture is highly performing
(2) BUdR killing (ER ≤ 0.1%) was more efficient than previosly reported with plant cells whenever restricted to a particular stage of the protoplast growth cycle(≤ the first division). These escaping rates were but somewhat higher than those described with animal cells, and this with no significant effect of light exposure for the lethal effect of BUdR. Actually, the intimate mechanism of action of BUdR in plant cells remains obscure;
(3) A larger number (13), and wider spectrum of aminoacid auxotrophs (6) were isolated in these experiments than previously reported with either TIP (total isolation procedure) (Sidorov and Maliga, 1982; King, 1983), or arsenate negative screening (Horsch and King, 1983). Thus his^-(3), ile^-(3), ile^-+val^-(3), leu^-(2), trp^-(1) and met^-(1) were recovered and demonstrated to be stable with respect to their requirements for more than two years of culture. The <u>his^-</u> and more particulary the <u>met</u> lines represent new classes of auxotrophs in higher plants.
(4) A comparison in terms of effeciency among these techniques is yet premature : enrichement techniques need further assessment and serious improvements. The recovery rates are still unsatisfactory, with as consequence too low induced mutation rates (≥ 10^{-7}, as compared to 10^{-4} and $1.4x10^{-4}$ obtained with TIP). Essentially this appears to be the consequence of both BUdR side-effects and suboptimal timing of dilution following BUdR incubation
(5) Further comparison of our data with those from the literature indicates that identical altered fuctions leading to aminoacid auxotrophy (such as <u>ile^-</u>

val⁻+ile⁻, trp⁻, his⁻) were repeateadly recovered irrespective of the method of selection, mutagen treatment, and plant species;
(6) Regenerated auxotrophs, exhibiting the same absolute requirement for the corresponding aminoacid(s) as the original cell lines, were obtained from most of the auxotrophs. We suspect that certain difficulties we encountered in regenerating the auxotrophic lines were the direct consequence of the BUdR treatment per se.
(7) Although rather similar in their response to deficiency, starvation conditions etc, the auxotrophs behaved differently in meiosis : NR⁻ (the control marker used), and met⁻ mutants could be transmited sexually, while his⁻ failed to do so. Much emphasis was put on the genetic analysis of fusion-complementing products or revertants, but the resulting picture was rather confusing. In other reported cases, ile⁻ from N. plumbaginifolia did not survive grafting, and regenerants from fusion products were sterile (Sidorov and Maliga, 1982). Furthermore, fusion between NR⁻ tobacco and nic⁻ H. muticus produced Hyoscyamus-like phenotypes but neither the NR⁻ nor the nic⁻ traits segregated out in the progeny analysis of the interspecific somatic hybrid plants (Potrykus et al, 1984).
(8) We could demonstrate that auxotrophy is compatible with normal developmental processes, although possible specific developmental implications remain to be substantiated. Further attempts, making use of anther and embryo culture will be reported elsewhere.
(9) Obtention of auxotrophic mutant plants in experimental systems with high protoplasting ability is essential for the development, understanding, and achievement of genetic manipulation in eukaryotes.

References :

Horsch R.B., J.King (1983) Planta 159 : 12-17
King P.G. (1983) IAPTC Newsletters 39 : 2-7
Potrykus I., J.Jia, G.B. Lazar, M. Saul (1984) Plant Cell Reports 3 : 68-71
Sidorov V.A., P. Maliga (1982) Molec. Gen. Genet. 186 : 328-332

SELECTABLE MARKERS AND THEIR USE IN THE GENETIC MANIPULATION OF PLANT CELLS : II SETTING UP CELLULAR EXPERIMENTAL SYSTEMS IN VIEW OF GENE TRANSFER.

Y. Famelaer, M. Heinderickx, I. Negrutiu, M. Jacobs.
Plantengenetica, I.M.B., 1640 St Genesius Rhode, Belgium.

Several types of biochemical mutants (resistant to inhibitory concentrations of aminoacid(s) or aminoacid analogs as well as deficient for enzyme acitivity) have been isolated and characterized biochemically and genetically (in N. plumbaginifolia and N. sylvestris). They are being used in the development and refinement of gene transfer techniques.

Three lines of investigation have been followed thus far.

(1) Studying the expression and selection conditions of putative transforants.

In reconstruction-simulation experiments we selected back wild type cells from varying dilutions (1:100 to 1:10,000) into a background of a constant concentration of nitrate reductase-less (NR^-) cells of N. plumbaginifolia with recovery rates above 50% plating efficiency.

Another approach consisted in selecting for revertants in protoplast populations isolated from auxotrophic plants, namely NR^- and his^- mutants. Both spontaneous and induced revertants were obtained, at reversion frequencies of approximately 10^{-7} and 10^{-6} respectively.

These experiments indicated that transformant recovery can be performed under reproducible and reliable conditions.

(2) Total DNA-mediated transformation.

Protoplasts from NR^- mutants were incubated in the presence of polycations with high molecular weight DNA extracted from donor wild type and NR^- (control) cells. Corrected phenotypes were recovered from both treatments at similar frequencies (wich were in average 10-fold higher than the spontaneous reversion rates and similar to those reported with animal cells. All 23 selected lines were stable, and exhibited lower specific NR activity than wild type controls. Genetic analysis of regenerated clones showed that in all tested cases the corrected phenotype was allelic to the wild type gene. We concluded on the mutagenic effects of the donor DNA as the most plausible and frequent mechanism of "transformation".

(3) Fusion of protoplast with an irradiated partner(γ-fusion).

The objective is to transfer small genetic attributes (fragments of chromosomes, or even a few genes) from the irradiated donor partner into a receptor deficient mutant(NR^- and trp^-). To avoid undesirable genetic variability, partners with balanced, euploid genomic constitution (leaf protoplasts) were chosen. Combination of heavy radiation effects and specific selection pressure which is applied after the first division of the culture, together with regeneration induction (i.e. developmental selection pressure), are being studied as possible means of forcing both gene transfer events and the sequential elimination of redundant genetic material. The role played by incompatibility mechanisms is assessed by using as donor partners protoplasts of N. sylvestris (albino), D. innoxia ($5MT^R$), and sugarbeet (habituated). The results show that corrected phenotypes occur in all combinations tested. Doses below 50 Krad produced but partial hybrids, and higher doses are presently tested.

VALINE-RESISTANT PLANTS FROM **NICOTIANA PLUMBAGINIFOLIA** MESOPHYLL PROTOPLASTS

A. MARION-POLL, C. MISSONIER, J. GOUJAUD, M. CABOCHE

1. INTRODUCTION

Valine-resistant plants have been obtained from mesophyll protoplasts of **Nicotiana tabacum** and **N. sylvestris** (1,2). Valine-resistance was recently used as a marker to develop reproducible methods for the isolation of tobacco valine-resistant plants **in vitro** (3). We report here preliminary results suggesting that valine resistance is a convenient genetic marker to study the efficiency of mutagenic treatments on **N. plumbaginifolia** protoplasts.

2. PROCEDURE

2.1. Material and methods

2.1.1. Plant material, media and selection conditions. Mesophyll protoplasts were prepared from haploid plants of **N. plumbaginifolia** and cultured in medium T_0 as previously described (4). One day after isolation, protoplasts were irradiated with a germicidal lamp (G30T8) at a incident dose rate of 12.5 erg/mm^2/sec. After 5 days in the dark, protoplast-derived cells were subcultured in medium C with 0.05 µM of naphthalene acetic acid at a density of $5x10^3$ cells/ml (5). Two weeks later, cells were plated at 10^3 cells/ml on modified medium C with 0.05 µM naphthalene acetic acid, 0 or 0.22 M mannitol, 5% agarose and 5 or 10 mM L-valine. After one month of culture on selective medium, surviving colonies were subcultured onto modified medium B (20 mM KNO_3 as sole nitrogen source, 14 mM saccharose and no mannitol) (4). Regenerated plantlets were propagated **in vitro** on the same medium and some of them were transferred to the greenhouse for subsequent protoplast isolation.

2.1.2. Characterization of selected clones. Protoplast-derived cells of the selected clones were subcultured in medium C with different valine concentrations at a density of 10^2 cells/ml. Resistance of seeds was tested on medium B containing 1 or 2 mM valine.

2.2. Results and discussion

Among the 98 presumed valine-resistant clones selected on medium C without mannitol (Table 1), about 50 have survived and regenerated; 9 were tested for resistance after transfer to the greenhouse. Protoplast-derived cells of 5 clones were resistant to valine (Fig. 1). Plants of these 9 clones could not be selfed but 4 of them have already been crossed with wild type plants; resistance test of the seeds are under progress. The presumed mutant clone obtained without irradiation (Table 1) was also transferred to the greenhouse and was selfed; seeds were resistant to 1 mM valine whereas wild type seeds were killed at the same valine concentration.

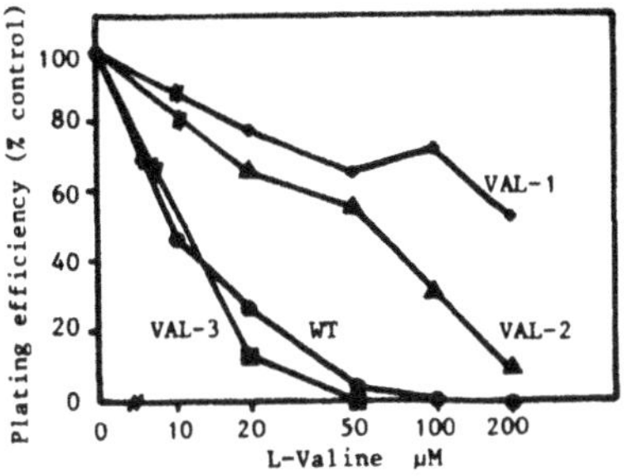

FIGURE 1. Valine resistance of protoplast-derived cells isolated from leaves of regenerated presumed mutant clones grown at low density (100 cells/ml).

Table 1. Effect of culture conditions on the recovery of valine-resistant colonies.

Protoplast treatment	No irradiation	UV irradiation	
Mannitol concentration in the selective medium	0	0	0.22 M
Number of dividing cells submitted to selection	$7x10^4$	$2.6x10^5$	$3x10^5$
Number of surviving colonies	1	98	10

REFERENCES

1. Bourgin JP. 1978. Mol. Gen. Genet. 161, 225-230.
2. Vunsh RD, Aviv, Galun E. 1982. Theor. Appl. Genet. 64, 51-58.
3. Grandbastien MA, Bourgin JP, Caboche M. Genetics (in press).
4. Bourgin JP, Chupeau Y, Missonier C. 1979. Physiol. Plant. 45, 288-292.
5. Muller JF, Missonier C, Caboche M. 1983. Physiol. Plant. 57, 35-41.

SYNCHRONIZATION, FLOW CYTOMETRIC SORTING AND MICROINJECTION STUDIES USING NICOTIANA PLUMBAGINIFOLIA CELLS AND PROTOPLASTS

H. VERHOEVEN, J. MOTTLEY, A. DE LAAT, K.J. PUITE, B. DE GROOT
Research Institute ITAL
P.O. Box 48, 6700 AA Wageningen, The Netherlands.

1. INTRODUCTION

Within the framework of our research which is directed towards chromosome mediated gene transfer in plant cells, making use of genetic complementation of nitrate reductase deficient (NR^-) cells of *Nicotiana plumbaginifolia*, attention has been paid to:

a. the synchronization of cell suspension cultures for the isolation of metaphase chromosomes,

b. sterile sorting using a flow cytometer. Leaf protoplasts have been sorted, cultured and regenerated to plants,

c. microinjection of protoplasts. Experience has been gained with the injection system using a fluorescent stain.

2. PROCEDURES AND RESULTS

2.1. Synchronization of cells

Use of hydroxyurea (HU) and colchicine which arrest cells at the G_1/S border and in metaphase, respectively, as described for *Haplopappus gracilis* (1) resulted in mitotic indices (MI) up to 30-40% in cell suspension cultures of the NR^- *N. plumbaginifolia* cells (NA36 line). However, severe clumping of the metaphase chromosomes was observed. This clumping also occurred in the presence of colchicine alone (0.05%, MI 30%). Treatment with HU alone resulted in many well-spread chromosomes but, because the cells are not arrested in metaphase, a high MI has not been obtained. HU (10 mM for 12 h) in combination with the microtubule poison griseofulvin (10^{-5} - 10^{-6} g/ml for 12 h) gave MI of up to 20% without visible chromosome clumping for both NA36 and wild type cells.

Chromosomes were isolated from wild type cells following the procedure for *H. gracilis* (1). Flow cytometric DNA distribution histograms, measured after purification of the chromosome suspension on a sucrose gradient, showed a series of peaks. Subsequent sorting showed that these peaks were present in an area of the histogram which contained an enriched fraction of chromosomes.

2.2. Flow cytometric sorting of protoplasts

The possibility of sterile sorting of chromosomes and protoplasts (2) using vital fluorescent stains or endogenous fluorescence has several applications, including some in the field of chromosome transfer studies.

Mesophyll protoplasts (diameter 20-50 μm) from *N. plumbaginifolia* wild type plants were sorted based on their chlorophyll autofluorescence. The 488 nm line of the argon-ion laser in the FACS IV cell sorter was used, while a long pass 620 nm emission filter was installed. The diameter of the nozzle tip orifice was 80 μm. Sterilization of the apparatus was achieved by overnight flushing with a mixture of 70% ethanol and 2% lyorthol.

Deflected droplets containing the sorted protoplasts were collected on a 0.5 M mannitol layer and after centrifuging, the pellet was transferred to culture medium. The percentage of dividing cells after 5 d of culture was 30% compared to 40% for the control and whole plants were obtained after 3½ months.

2.3. Microinjection of protoplasts

Microinjection instrumentation was constructed based on that of Ansorge (3). Protoplasts of *N. plumbaginifolia*, immobilized in a thin layer of agarose, were microinjected with the fluorescent dye lucifer yellow and with DNA. So far cell division is poor. Injection of chromosomes is currently under investigation.

REFERENCES

1. de Laat AMM, Blaas J. 1984. Flow-cytometric characterization and sorting of plant chromosomes. Theor. Appl. Genet. 67, 463-467.
2. Harkins KR, Galbraith DW. 1984. Flow sorting and culture of plant protoplasts. Physiol. Plant. 60, 43-52.
3. Ansorge W. 1982. Improved system for capillary microinjection into living cells. Exp. Cell Res. 140, 31-37.

CHARACTERIZATION OF THE MOLECULAR LESION RESPONSIBLE FOR TEMPERATURE SENSITIVE EMBRYOGENESIS IN A CARROT CELL MUTANT

F.LOSCHIAVO, G. GIULIANO, M.TERZI

The application of genetic engineering techniques to plant species is often prevented by lack of regeneration. An understanding of the mechanisms regulating the process is therefore a priority task. In a model system, like carrot, several mutants showing temperature-sensitive embryogenesis (ts-emb$^-$) have been isolated (1,2,3) with the aim of identifying and dissecting developmental processes occurring both somatically and zygotically. One of these mutants, ts59 (1) has been characterized.

ts59 at the non-permissive temperature (32°C) is blocked at an early globular stage. The mutation is stable after several cycles of dedifferentiation-regeneration and is likely to be dominant as it does not complement -in somatic hybrids- lines complementing among themselves.

When carrot cells are exposed to a temperature of 32°C there is a transient synthesis of heat shock proteins (hsp) and the normal pattern is restored in less than 24 hours (4). By comparing the hsp pattern -after two-dimensional polyacrylamide gel electrophoresis of wild type (HA line) and ts59 exposed to 32°C, it is easy to identify -in the region of 38 and 42-45 kD- a string of spots in wild type that appear as single spots in ts59.

This difference disappeared when the electropherograms made after in vitro translation of isolated mRNA polyA$^+$ were compared single spots appearing in both cases. This suggests a post-translational modification of the wild type products which cannot occur in ts59.

This secondary modification has been, in subsequent experiments, identified as a phosphorylation.

Embryo-specific hsp have been described in carrot cell lines (4) and analyses of these proteins in polyacrylamide gel electrophoresis are now in progress with the aim to understand the stage specific sensitivity of ts59 during early embryogenesis.

ACKNOWLEDGEMENT

F.L.S.carried out most of her experiments while on leave at the University of California (Dr.Z.R.Sung's laboratory).

This work was supported by E.E.C. grant GBI-6-031-I(S) and by CNR, P.F. "Ingegneria Genetica".

REFERENCES

1. Breton A.M., Sung Z.R.(1982) Dev.Biol.90:58
2. Terzi M.,Giuliano G.,Lo Schiavo F.,Nuti Ronchi V.(1982).In: "Embryonic Development"(M.M.Burger ed.)part B p;521 Alan R.Liss, New York.
3. Giuliano G., Lo Schiavo F., Terzi M. (1984) Theor.Appl.Genet. 67:179
4. Pitto L., Lo Schiavo F., Giuliano G., Terzi M. (1983) Plant Mol.Biol.2:231

PARALLEL SESSION I

P A R A L L E L   S E S S I O N   I

Workshop on cytoplasmic male sterility

Chairman : Prof. D.R. Davies

Structure and function of mitochondrial genomes

M. Briquet : The mitochondrial translation products of plants

B.M. Hansen : Mitochondrial plasmids in sugar beet

A.J. Kool : Mitochondrial genetics in Petunia

K. Esser : Molecular cloning with mitochondrial DNA

F. Quétier : Relation between mitochondria and cms.

Isolation or induction of mitochondrial mutations

A. Cornu : Induction and isolation of mitochondrial variants

A. Berville : A case of natural mt-DNA variability in the progenies of a Peruvian maize

WORKSHOP SESSION ON CYTOPLASMIC MALE STERILITY

Copenhagen, 9-10 October 1984

Report by D.R. Davies

The continuing success of the agricultural and horticultural sector of the Community's activities is dependent on the provision of a succession of new varieties of plants which are resistant to pests and diseases and have improved yield and quality. For many of the more important crops these new varieties are produced by intercrossing selected parental lines to produce hybrid forms which are then grown commercially. The large scale automated production of these hybrid varieties is often achieved by incorporating cytoplasmic male sterility (cms) into one parent; this ensures that only hybrid seed is produced.

It is necessary and important to understand the basis of cms for several reasons; for example, to speed up the production of new hybrid varieties, plant breeders must have easier methods either of introducing cms into a given strain, or some means of interconverting cms and fertile forms. In addition, if cms can be produced in, or introduced into, crop species in which it cannot be found at present then this could allow the production of Fl hybrids in those crops. A further reason for studying the molecular basis of cms is that there is good evidence in maize that a particular form of cms and susceptibility to blight are closely linked traits. While there is considerable circumstantial evidence to suggest that cms is encoded by genetic information within the mitochondrion, we have little understanding of the phenomenon and the further exploitation of cms will have to await much more detailed elucidation of its molecular basis. The workshop held in Copenhagen considered many aspects of cms and of mitochondrial molecular biology and while many achievements could be identified it also hightlighted the many areas that demand much further investigation.

The evidence adduced in support of the claim that the genetic defect responsible for cms is to be found in the mitochondrial genome, is based mainly on studies of the S form cms in maize. Even here the precise defect has not been characterised. The large size and complexity of the organisation of the plant mitochondrial genome is one of many factors mitigating against an early resolution of the problem. This was highlighted in the workshop. Data were presented to show the range of techniques utilised to characterise these genomes, the wide range of genome length which exists in some plant species, the problem of maintaining the fidelity of their conformation during extraction, and the possible interrelationships of the different genomes. The existence of many different repeated sequences within the genome was demonstrated and these could allow particular kinds of recombination events to occur which in turn could generate a spectrum of different sized molecules within the mitochondrion.

The existence of mitochondrial plasmids - both linear and supercoiled circles (CCC) is widely recognised, and while differences exist in the profile of plasmids in cms and fertile forms in many species, no definitive causal relationship has yet been shown between the plasmids and cms. In the case of sugar beet, three of the CCC plasmids have been sequenced; they show no homology to the main mitochondrial genome but do show some homology with the nuclear genome. The implications of the homology between the plasmid and the nucleus awaits clarification.

There are more examples emerging of in vitro production of variant polypeptides by the mitochondrial genomes of cms forms. Their interrelationship with polypeptides of fertile forms however remains to be established, as does their possible role in the cms syndrome. The availability of probes produced from yeast mitochondrial genomes will help in studying the comparative mitochondrial functions of fertile and cms forms. In this and other respects the work on yeast and other fungi is providing useful techniques, information and probes for those working on plants. Differences in the respiratory activities and in the energy utilisation of mitochondria from cms and fertile forms are now beginning to emerge.

The induction and isolation of mitochondrial mutations is a particular problem, although there are indications that variant forms of cms can and will be generated. The use of protoplast fusion to derive new combinations of mitochondria and nuclei is being actively pursued; there is a sound basis for such studies in the classical evidence that novel combinations of nuclei and cytoplasm produced by wide sexual crosses can generate cms forms. Again there is evidence that by growing plant cells in tissue culture, changes can be generated in fertility and in the nuclease digest pattern of mitochondrial DNA. When tissue culture is combined with the use of mutagens or antibiotics this may further enhance the yield of mitochondrial variants. The extent to which the mitochondrial genome can be destabilised when introduced into a particular nuclear genetic background remains to be established, but there are clear indications in maize that particular combinations of nuclei and mitochondria lead to significant changes in the latter.

There is such a dearth of information on the plant mitochondrial genome - on its structure and its function, that all studies related to these topics need urgent support; until much more information has been accumulated on basic aspects related to these two fundamental issues it will not be possible to make the necessary progress on the problem of the induction, control and manipulation of cms in crop plants. Without question, genetic engineering will ultimately allow the ready manipulation of cms in plants, and agriculture and horticulture will benefit accordingly.

PARALLEL SESSION II

PARALLEL SESSION II

Workshop on symbiotic genes

Chairman : Dr. J. Tempé

Structure and function of symbiotic genes in the legume-Rhizobium symbiosis :

A.W.B. Johnston : Nitrogen fixation in legumes, host-range genes

A. Pühler : Fix and Nif genes in Rhizobia

K.A. Marcker : Plant genes involved in symbiosis

Other symbiotic associations showing plant growth stimulating properties

F. Meinhardt : Plasmids in mycorrhizal fungi

J. Marugg : Siderophore effect in *Pseudomonas putida*

Agrobacterium as a symbiont ?

M. Van Montagu : Virulence in *Agrobacterium*

J. Tempé : The opine concept in *Agrobacterium* and *Rhizobium*

WORKSHOP SESSION ON SYMBIOTIC GENES

Copenhagen, 9-10 October 1984

Report by J. Tempé

The word symbiosis is generally used to describe an association between organisms which benefits both partners. Originally this word had a wider sense. It was coined by De Bary in 1879 to describe "the living together of dissimilar organisms". De Bary distinguished between antagonistic and mutualistic symbiosis. Because there is no clear border-line separating these two forms of symbiosis and because it is obvious that many concepts developed in the study of one of them may be relevant to both, it was decided to come back, at least for the duration of the workshop, to the older definition.

Acquiring the ability to control and manipulate interactions between plants and microorganisms would have important economic consequence and could profoundly modify the conditions of agricultural production. Although some results of economic significance have already been obtained from studies on symbiotic nitrogen fixation (inoculation of legume crops with *Rhizobium sp*. or reforestation schemes involving trees able to form actinorhizae with *Frankia sp*.) and on crown gall (biological control of crown gall) it is perfectly clear that these are minor achievements in comparison to what can be expected. However, it is clear that, except for straightforward applications, the most significant developments will be based on a thorough knowledge at the molecular level of interactions between plants and microorganisms.

The laboratories involved in molecular studies on the subject within the framework of the CEC programme are among the best in Europe and have achieved worldwide recognition.

All the topics discussed raised much interest and, even if the workshop had been entirely devoted to *Rhizobium* - legume symbiosis, there would not have been enough time for discussion. This field of investigation is well represented among the contractant laboratories and studies at the fundamental level are in rapid progress. But owing to the complexity of the system, applications have still to be developed. Most laboratories concentrate on bacterial genes involved in the symbiosis. Advances have been made in the identification of genes involved in nitrogen fixation as well as of those involved in nodule formation. Knowledge gained here should ultimately result in the creation of more competitive and more efficient bacterial strains. Despite the fact that fewer groups are involved in the study of legume symbiotic genes, progress is rapid. The transcription of the soybean leghemoglobin genes during nodule development has been investigated in detail. Other nodule-specific plant genes (nodulin genes) are being studied.

Although studies on mycorrhizal associations have been going on for quite a long time, it is only recently that they have been envisaged at molecular level. For the time being, research is still at an early stage and the primary objective is to identify the genetic structures which code for symbiotic functions.

The plant growth promoting effect of non pathogenic fluorescent Pseudomonas has been investigated for many years in several laboratories in the world. Positive results are often not reproducible and efforts to use this growth promoting effect for improving crop yields have been unrewarding. It is hoped that the molecular studies on this interaction which have recently been initiated by the contractants will solve some of the problems encountered with the development of practical applications.

One of the speakers stated that bacteria "talk" to plants during legume symbiosis interactions. The statement also applies to these interactions between Agrobacterium and plant cells wich result in the formation of hyperplastic overgrowths called gall or hairy root. Here it is known that the message to the plant is a DNA sequence, the T-DNA, transferred from Agrobacterium Ti or Ri pathogenic plasmids to the plant cell. It is for this system that symbiotic genes (in the wider sense of the word) can be most clearly defined. Two types of function are involved in the interaction. Virulence functions are responsible for the transfer and integration of the T-DNA in the plant genome; they themselves do not become integrated. T-DNA functions expressed in the plant cells appear to be specifically geared to provide favorable nutritional conditions for the pathogen. The first of these concerns the synthesis of opines, specific growth substrates for Agrobacterium strains carrying Ti or Ri plasmids. The second type of T-DNA function specifies plant cell proliferation resulting in crown gall or hairy root tumors. It has been proposed that this proliferation which results in continuously increasing mass of opine producing cells, enhances the positive selection for Ti or Ri plasmids provided by opines released from transformed cells. The discovery of opine-like compounds in some legume nodules has been reported, but its significance is not understood. Apart from the point of view of symbiosis, which was the main point in discussion, the use of Ti or Ri plasmid-derived vectors for plant genetic engineering, stimulates much interest. In this respect the reports that opines have been found in tissues from monocots inoculated with Agrobacterium tumefaciens constitute an important breakthrough.

As has been said above, most of the applications that will result from manipulating interactions between plants and microorganisms are still to come and, in the present state of knowledge, detailed predictions would be unrealistic. However, some schemes are being studied; although it is difficult, for lack of knowledge, to describe them with accuracy one can be reasonably sure that they will ultimately reach their objectives if sufficient efforts are invested for the solution of pending problems. Thus, the development of more competitive strains of Rhizobium essentially depends upon an accurate description of nodulation at the moledular level. When the regulation of symbiotic nitrogen fixation is better understood it should also be possible to engineer Rhizobium strains that will fix nitrogen more efficiently. One of the first applications of plant genetic engineering to the improvement of agricultural crops will be the introduction of herbicide resistance into cultivated varieties. More ambitious achievements will result, at a later stage, from a better understanding of functions in the plant genome. Except for symbiotic nitrogen fixation in legume nodules, crown gall and hairy root systems, very little is known at present of the interactions between plants and soil microorganisms. Studies which are just beginning at the molecular level will contribute to the understanding of several systems and, thereafter, to the development of new mechanisms of biological control. The creation of artificial symbiosis could be another fallout from research in this direction.

LIST OF PARTICIPANTS

Name	Institute	Origin	Page
Bazetoux, S.	Lab. de Biologie Moléculaire Végétale Université Paris-Sud	Orsay	no poster
Beier, H.	Abt. Pflanzliche Zellkulturen G.B.F.	Braunschweig	124
Berlin, J.	Abt. Pflanzliche Zellkulturen G.B.F.	Braunschweig	124
Berville, A.	Station d'Amélioration des Plantes / I.N.R.A.	Dijon	130 142
Boistard, P.	Lab. Biologie Moléculaire des Relations Plantes - Micro-organismes / INRA-CNRS	Castanet	48
Boulter, D.	Dept. of Botany University of Durham	Durham	32 100 102
Bremenkamp, M.	Istituto Sperimentale per la Cerealicoltura	Bergamo	116
Briquet, M.	Lab. d'Etude de l'Hérédité Cytoplasmique UCL	Louvain	136 146 148
Broughton, W.J.	Max-Planck-Institut	Köln	56
Caboche, M.	Laboratoire de Biologie Cellulaire / C.N.R.A.	Versailles	174
Cardarelli, M.	Dipartimento Genetica e Biologia Molecolare Università di Roma	Roma	78
Casey, R.	John Innes Institute	Norwich	104
Charbonnier, M.	Lab. d'Etude de l'Hérédité Cytoplasmique UCL	Louvain	148
Chojecki, A.	Dept. of Physiology Carlsberg Laboratory	Copenhagen	114

Name	Institute	Origin	Page
Chupeau, Y. . . .	Lab. de Biologie Cellulaire INRA	Versailles . . .	168
Cornu, A.	Station d'Amélioration des Plantes / I.N.R.A.	Dijon	142
Costantino, P. .	Dipartimento Genetica e Biologia Molecolare Università di Roma	Roma	78
Dale, P.J. . . .	Welsh Plant Breeding Station	Aberystwyth . .	162 164
Davies, D.R. . .	John Innes Institute	Norwich	144
Dean, P.D.G. . .	Agricultural Genetics Comp. .	Cambridge .	no poster
De Groot, B. . .	Research Institute ITAL . . .	Wageningen . . .	176
Del Giudice, L. .	Ist. Internazionale di Genetica e Biofisica / C N R . .	Napoli	120
Deshayes, A. . .	Laboratoire de Biologie Cellulaire / C.N.R.A.	Versailles .	no poster
Di Fonzo, N. . .	Istituto Sperimentale per la Cerealicoltura	Bergamo	116
Dowling, D.N. . .	Max-Planck-Institut	Köln	56
Durrens, P. . . .	Dept. of Biology University of Essex	Colchester . . .	94
Elmerich, C. . .	Unité de Physiologie Cellulaire / Institut Pasteur . .	Paris	34 46
Esser, K.	Lehrstuhl für Allgemeine Botanik / Ruhr-Universität .	Bochum	38 92
Fecker, L. . . .	Abt. Pflanzliche Zellkulturen G.B.F.	Braunschweig . .	124
Fedoroff, N. . .	Dept.of Embryology / Carnegie Institution of Washington . .	Baltimore . . .	30
Foulger, D. . . .	Biochemistry Dept. Rothamsted Exp. Station . .	Harpenden . . .	154
Gausing, K. . . .	Dept. of Molecular Biology University of Aarhus	Aarhus	126

Name	Institute	Origin	Page
Gebhardt, C. . .	Biochemistry Dept. Rothamsted Exp. Station . .	Harpenden . . .	68
Giese, H.	Botanisches Institut der Universität Kiel	Kiel	110 128
Gilmour, M. . . .	Dept. of Botany University of Nottingham . .	Nottingham . . .	158 160
Goblet, J.P. . .	Lab. d'Etude de l'Hérédité Cytoplasmique UCL	Louvain	136
Götz, R.	John Innes Institute	Norwich	54
Grimm, C.	Inst. of Molecular Biology Research Centre of Crete . .	Heraklion . . .	98
Hale, A.D. . . .	Welsh Plant Breeding Station	Aberystwyth . .	164
Hensgens, L. . .	Ist. di Anatomia Comparata Biologia Generale e Genetica.	Florence	86
Hofstadt, G. . .	Dept. of Molecular Cell Biology / Univ. of Utrecht .	Utrecht . .	no poster
Hoge, J.	Dept. of Plant Molecular Biology / S.U.L.	Leiden	74 76
Hooymans, C.M. .	Dept. of Genetics A.U.W.	Wageningen . . .	166
Huguet, T. . . .	Lab. Biologie Moléculaire des Relations Plantes - Micro-organismes / INRA-CNRS . . .	Castanet	48
Jacobs, M. . . .	Inst. of Molecular Biology Free University of Brussels .	Brussels	170 172
Johnston, A.W. .	John Innes Institute	Norwich	52 54
Jonard, G. . . .	Inst. de Biologie Moléculaire et Cellulaire / C.N.R.S. . .	Strasbourg . . .	90
Jones, M.G.K. . .	Biochemistry Dept. Rothamsted Exp. St.	Harpenden . . .	154 156
Jubier, M.F. . .	Lab. de Biologie Moléculaire Végétale Université Paris-Sud	Orsay	132 134

Name	Institute	Origin	Page
Kool, A.J.	Dept. of Genetics Free Univ. Amsterdam	Amsterdam	138 140
Kreiberg, J.	Dept. of Molecular Biology University of Aarhus	Aarhus	126
Kreis, M.	Biochemistry Dept. Rothamsted Exp. St.	Harpenden	106
Krens, F.	Dept. of Plant Molecular Biology / S.U.L.	Leiden	74
Lo Schiavo, F.	Istituto Mutagenesi e Differenziamento / C.N.R.	Pisa	178
Lotz, W.	Lehrstuhl für Mikrobiologie Universität Erlangen	Erlangen	62 64
Magnien, E.	Commission of the European Communities	Brussels	no poster
Manian, S.	Universität Bielefeld	Bielefeld	60
Marcker, K.A.	Dept. of Molecular Biology University of Aarhus	Aarhus	70
Marotta, M.	Istituto Sperimentale per la Cerealicoltura	Bergamo	116
Meinhardt, F.	Lehrstuhl für Allgemeine Botanik / Ruhr-Universität	Bochum	92
Merrick, M.J.	A.R.C. University of Sussex	Brighton	36 44
Mol, J.N.M.	Dept. of Genetics Free Univ. Amsterdam	Amsterdam	122
Müller-Gensert	Max-Planck-Institut	Köln	152
Negrutiu, I.	Inst. of Molecular Biology Free University of Brussels	Brussels	170 172
Nettancourt,D.de	Commission of the European Communities	Brussels	no poster
Nijkamp, H.J.J.	Dept. of Genetics Free Univ. Amsterdam.	Amsterdam	138 140

Name	Institute	Origin	Page
Schieder, O.	Max-Planck-Institut	Köln	26 152
Selbitschka, W.	Lehrstuhl für Mikrobiologie Universität Erlangen	Erlangen	no poster
Simpson/ Cameron-Mills, V.	Dept. of Physiology Carlsberg Laboratory	Copenhagen	112
Sotak, P.	Dept. of Botany University of Nottingham	Nottingham	158
Sybenga, J.	Dept. of Genetics Agric. Univ. Wageningen	Wageningen	166
Tempé, J.	Institut de Microbiologie C.N.R.S.	Orsay	82
Tempelaar, M.J.	Biochemistry Dept. Rothamsted Exp. St.	Harpenden	156
Terzi, M.	Istituto Mutagenesi e Differenziamento / C.N.R.	Pisa	178
Thompson, R.D.	Plant Breeding Institute	Cambridge	108
Tichy, H.V.	Lehrstuhl für Mikrobiologie Universität Erlangen	Erlangen	62 64
Van Emmelo, J.	Lab. of Molecular Biology Rijksuniversiteit Gent	Gent	88
Van Montagu, M.	Laboratorium voor Genetica Rijksuniversiteit Gent	Gent	72
Veltkamp, E.	Dept. of Genetics Free Univ. Amsterdam	Amsterdam	122
Verhoeven, H.	Research Institute ITAL	Wageningen	176
Viotti, A.	Istituto Biosintesi Vegetali CNR Milano	Milano	118
Walgate, R.	Nature Magazine	London	no poster
Weisbeek, P.J.	Dept. of Molecular Cell Biology / Univ. of Utrecht	Utrecht	96

9 789024 731312

GPSR Compliance

The European Union's (EU) General Product Safety Regulation (GPSR) is a set of rules that requires consumer products to be safe and our obligations to ensure this.

If you have any concerns about our products, you can contact us on ProductSafety@springernature.com

In case Publisher is established outside the EU, the EU authorized representative is:

Springer Nature Customer Service Center GmbH
Europaplatz 3
69115 Heidelberg, Germany

Batch number: 10809441

Printed by Printforce, the Netherlands